THE INTERMITTENT FASTING REVOLUTION

THE INTERMITTENT FASTING REVOLUTION

The Science of Optimizing Health and Enhancing Performance

MARK P. MATTSON

The MIT Press
Cambridge, Massachusetts
London, England

First MIT Press paperback edition, 2023
© 2022 Massachusetts Institute of Technology

The MIT Press would like to thank the anonymous peer reviewers who
provided comments on drafts of this book. The generous work of academic experts is essential for establishing the authority and quality of our
publications. We acknowledge with gratitude the contributions of these
otherwise uncredited readers.

This book was set in Adobe Garamond Pro and Berthold Akzidenz
Grotesk by Westchester Publishing Services. Printed and bound in the
United States of America.

Library of Congress Cataloging-in-Publication Data

Names: Mattson, Mark Paul, author.
Title: The intermittent fasting revolution : the science of optimizing
   health and enhancing performance / Mark P. Mattson.
Description: Cambridge, Massachusetts : The MIT Press, 2021. |
   Includes bibliographical references and index.
Identifiers: LCCN 2021001823 | ISBN 9780262046404 (hardcover)—
   9780262545983 (paperback)
Subjects: LCSH: Intermittent fasting. | Longevity. | Nutrition.
Classification: LCC RM222.2 .M3792 2021 | DDC 613.2/5—dc23
LC record available at https://lccn.loc.gov/2021001823

10 9 8 7 6 5 4 3

To my loving wife Joanne, son Elliot, and daughter Emma, who have helped me immensely through thick and thin

Contents

Preface

Only recently in their evolutionary history have humans had the luxury of eating breakfast, lunch, and dinner without having to forage or hunt. Animals living in the wild, including humans, evolved in environments where food was sparse and meals were obtained only intermittently. This is perhaps best illustrated by predators such as wolves and lions, who commonly kill a prey animal only once every week or so. When the predators are in a food-deprived or fasted state, they must be able to determine where they are most likely to find prey and then expend the physical effort required to catch and kill their prey. Individuals whose brains and bodies functioned optimally when in a fasted state were those that survived and passed their genes on to the next generation. As you will discover in this book, evolution has "sculpted" our cells and organ systems such that they respond to intermittent fasting in ways that enable them to function optimally.

Intermittent fasting is not a diet. A diet is what is eaten and how much is eaten. In contrast, intermittent fasting is an eating pattern that includes frequent periods of time with little or negligible amounts of food of sufficient duration to cause "fat burning." You are in a fasted state when all of the energy (glucose) stored in your liver has been depleted and fats released from your fat cells are then converted to ketones. In humans, the liver contains enough glucose to last no more than 12 hours, and ketone levels in the blood then begin to rise. Ketones are used as a fuel for nerve and muscle cells during fasting, and they also stimulate these cells in ways that enable them to resist stress and combat disease. Three of the most popular intermittent fasting eating patterns are daily fasts of 16 to 20 hours, which is often referred to as "daily time-restricted eating"; fasting two days every week, which has been dubbed "5:2 intermittent fasting"; and fasting five days in a row every month.

The potential value of lifestyles that incorporate intermittent fasting and applications of intermittent fasting in medical practice are vast. It comes at no cost and can, in fact, save both money and time. The last chapter of this book provides practical advice for adopting an intermittent fasting eating pattern. It's actually quite easy for many people. It will, however, take up to a month for the brain and body to fully adjust to an intermittent fasting eating pattern.

During the past three decades, I and scores of scientists who have worked in my laboratory have discovered how cells in the brain, cardiovascular system, muscles, and other organs respond to intermittent fasting in ways that improve their performance and resilience. I have written this book to share with the reader what I have learned about how the brain and body respond to intermittent fasting in ways that protect them against a wide range of diseases.

In 1935, a nutrition researcher at Cornell University named Clive McCay discovered that when he reduced the amount of food that he gave rats each day throughout their adult life (30 percent less food than they would eat when food was always available), they lived significantly longer. Because at that time there was virtually no ongoing research on aging, it would not be until the 1980s that efforts began to understand why what was then called "caloric restriction" slows down the aging process. I became aware of the "anti-aging" effect of caloric restriction in the early 1990s, and because aging is the major risk factor for Alzheimer's and Parkinson's diseases and stroke, I recruited several postdoc scientists to perform some simply designed yet profoundly revealing experiments. We found that when rats or mice were maintained on an every-other-day fasting regimen, neurons in their brains were resistant to dysfunction and degeneration in experimental models of Alzheimer's, Parkinson's, and stroke.

During the past 20 years, my research has elucidated what happens in the nerve cell circuits in the brain that explains the remarkable beneficial effects of intermittent fasting. We also discovered very profound improvements in blood sugar regulation, cardiovascular stress resistance, and physical endurance in animals on intermittent fasting. I then began collaborating with physicians and dieticians in studies of intermittent fasting in humans. These human studies showed that intermittent fasting improves body composition (reduced fat) and glucose regulation and reduces inflammation and oxidative stress.

In 2019, my research collaborator and friend Rafa de Cabo and I were invited to write a review article on the effects of intermittent fasting on health, aging, and disease for the *New England Journal of Medicine*. The journal's editors felt it was time for such an article for two reasons. First, physicians are increasingly being asked about intermittent fasting by their patients, and yet many of them are not familiar with the state of research on intermittent fasting. Second, a sufficient number of randomized controlled trials of intermittent fasting in humans have clearly shown its effectiveness for losing weight and improving risk factors for diabetes and cardiovascular disease. The National Institutes of Health are currently encouraging clinical investigators to apply for grants to study the potential benefits of intermittent fasting for people at risk for or suffering from a range of chronic diseases.

There is now sufficient evidence from human studies to justify the prescription of intermittent fasting for people with obesity or type 2 diabetes or both, and evidence is emerging that intermittent fasting may prove useful in the treatment of many different medical conditions, including cancers and some neurological disorders. As I write, we are in the midst of the COVID-19 pandemic. Age, diabetes, and obesity are the major risk factors for poor outcomes and death among people who are infected. Inasmuch as intermittent fasting can counteract these risk factors, it seems likely to provide a hedge against the complications of COVID-19, although this remains to be determined.

Unfortunately, major obstacles inherent in our profit-driven food, pharmaceutical, and health industries have the potential to prevent incorporation of intermittent fasting interventions for disease prevention and treatment in our health care system. The ubiquitous marketing of processed foods and drugs discourages healthy lifestyles, and so intermittent fasting is a lifestyle change that has the potential to decrease those industries' bottom line. Chapters 8 and 9 describe how this unfortunate situation might be reversed and how intermittent fasting can play a major role in that reversal.

INTRODUCTION

When you eat a meal that contains carbohydrates, the energy from the food is first stored as glucose in your liver. When the liver is full, then energy is stored in the fat depots of the body. The liver can store about 400–500 calories of glucose, whereas fat can store tens of thousands of calories. The energy from fat can enable a person to fast for many weeks or months. Intermittent fasting is an eating pattern that includes frequent fasting periods of sufficient time duration to deplete liver glucose stores and so switch the body to using the ketones produced from fats. This "metabolic switch" from glucose to ketones requires at least 12 hours of fasting. Figure I.1 shows examples of what happens to glucose and ketone levels in the blood with two intermittent fasting eating patterns compared to the typical American eating pattern of three meals plus an evening snack every day.

The lay public and media outlets often consider intermittent fasting a type of diet, but intermittent fasting is in fact not a diet—it is an eating pattern. A diet is defined by what is eaten and how much of those foods are eaten. In contrast, intermittent fasting concerns when and how often food is eaten. This is not to say that intermittent fasting is a license to

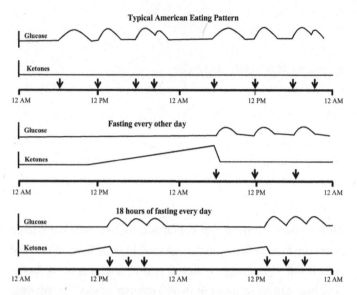

Figure I.1

Diagram showing changes in blood glucose and ketone levels during a two-day period in people with three different eating patterns. A typical American eating pattern of breakfast, lunch, and dinner plus an evening snack results in an elevation of glucose levels every time the person eats. Ketone levels do not rise because liver glucose stores are replenished every time the person eats. In the case of a person who fasts every other day, glucose levels will remain low, and ketone levels will rise on the fasting days. In the case of a person who fasts for 18 hours every day by consuming food only between noon and 6:00 p.m., ketone levels will rise during the morning hours every day.

eat junk food. It is not. Nutritionists and scientists generally agree that healthy diets include a variety of vegetables, fruits, nuts, whole grains, and healthy meats such as fish. A healthy diet excludes sugars (particularly fructose) and trans fats and limits saturated fats and red meats. A healthy diet also limits daily calorie intake to an amount that places one's body mass index within what is considered the optimal range of 18 to 23.

Several different regimens of intermittent fasting have become popular. Perhaps the simplest to adopt is daily time-restricted eating that involves compressing one's eating time window to 6–8 hours every day. A person on this intermittent fasting regimen fasts for 16–18 hours every day, which means that she goes at least 4 to 6 hours in "fat-burning" mode. Another intermittent fasting regimen is called "5:2 intermittent fasting" and involves a typical three meals per day five days each week but eating no more than 600 calories two days each week. A person following 5:2 intermittent fasting will deplete her liver glucose stores and elevate ketone levels on the two 600-calorie days. Some human studies of intermittent fasting have used an every-other-day modified fasting regimen in which 300–400 calorie days are alternated with unrestricted eating days. Another approach to intermittent fasting is to consume only about 700 calories on five consecutive days every month. Among these different intermittent fasting eating patterns, daily time-restricted eating is perhaps the easiest to adopt because it becomes part of one's daily routine. There have been no direct comparisons of these different intermittent fasting approaches with regard to their effects on health, but each has been shown to improve health indicators when compared to a breakfast, lunch, dinner, and snack eating pattern.

Intermittent fasting eating patterns often result in over-all reduction in weekly or monthly calorie intake. Therefore, people who are overweight find that intermittent fasting helps them get their weight down and keep it down. But an increasing number of studies are showing that at least some of the effects of intermittent fasting on health cannot be accounted for by weight loss alone. These effects include improved glucose regulation and blood lipid levels as well as a reduction in abdominal fat. Studies in humans have shown that intermittent fasting is effective in enabling weight loss and can improve health indicators in ways that suggest protection against diabetes, cardiovascular disease, and cancers. The brain can also benefit from intermittent fasting by reducing levels of anxiety, improving learning and memory, and protecting against common neurological disorders such as Alzheimer's disease and stroke.

Chapter 1 provides evolutionary and historical perspectives on fasting. When considered in the light of evolution, it becomes obvious that intermittent fasting eating patterns are normal, whereas three meals plus snacks are abnormal. During evolution, individuals whose brains and bodies functioned very well when they were in a food-deprived state were those who survived and passed their genes on to the next generation. We now know that many of those genes promote optimal performance and resilience under conditions of stress or disease. Chapters 2, 3, and 4 highlight some of these genes and how they respond to intermittent fasting in ways that slow aging, reduce the risk for many diseases, and improve brain and body performance. Chapter 5 recounts the development of a dietary "ketone ester" for the purposes of improving physical and mental performance and for treating Alzheimer's and Parkinson's diseases. Chapter 6 summarizes key features of a healthy diet and describes how some chemicals present in vegetables and fruits are beneficial for the surprising reason that they impose a mild stress on cells in the body and brain. Chapter 7 is intended for parents and pediatricians and focuses on the implications of intermittent fasting research findings for the health of children. The impact of parents' metabolic health on the risk for autism spectrum disorders in their children is highlighted. Chapter 8 points out the fact that there are profit-driven forces such as the pharmaceutical and processed-food industries that discourage healthy lifestyles that include intermittent fasting and exercise. Their bottom line is bolstered by encouraging people to eat unhealthy foods and by letting people get sick and then treating them with a drug to mitigate their symptoms. The book concludes with practical advice for people who would like to incorporate intermittent fasting into their lifestyle and for physicians who can include intermittent fasting in the care of their patients.

Here are several of the take-home messages to keep in mind as you read this book:

—*Because food is sparsely distributed, animals in the wild and our human ancestors prior to the Agricultural Revolution consumed food intermittently and in competition with others.*

—*Even the most advanced capabilities of the human brain (imagination, creativity, language, and magical thinking) evolved as adaptations to overcome food scarcity.*

—*A lifestyle characterized by three meals plus snacks every day and negligible exercise results in suboptimal brain function and increases the risk of major neurodegenerative and psychiatric disorders.*

—*Switching between time periods of negative energy balance (short fasts and/or exercise) and positive energy balance (eating and resting) can optimize general health and brain health.*

—*Health benefits of intermittent fasting go beyond those achieved with weight loss without intermittent fasting.*

—*With fasting and extended exercise, liver glycogen stores are depleted, and ketones are produced from fat-cell-derived fatty acids. This metabolic switch in cellular fuel source is accompanied by molecular adaptations of cells throughout the body and brain that enhance their functionality and bolster their resistance to stress, injury, and disease.*

—*By providing an alternative energy source and activating signaling pathways involved in neuroplasticity and cellular stress resistance, the ketone BHB (beta-hydroxybutyrate) plays a particularly important role in adaptations of nerve cell networks in the brain to fasting and exercise.*

—*The energetic challenges of fasting and exercise engage adaptive cellular stress-response signaling pathways in neurons involving proteins that bolster their ability to prevent and repair damage.*

—*Intermittent fasting slows aging, reduces inflammation, improves glucose regulation, lowers blood pressure, facilitates fat loss, and may reduce the risk of diabetes, heart disease, and cancers.*

—*Many different intermittent fasting regimens are likely to improve health, so individuals may choose one that suits their particular daily and weekly schedules.*

—*Intermittent fasting is not recommended for very young children, elderly frail persons, or people with an eating disorder. People with diabetes who are taking insulin or other glucose-lowering drugs should consult their physician before initiating intermittent fasting.*

1 FOOD SCARCITY "SCULPTED" THE HUMAN BRAIN AND BODY

Before we delve into what is known about the effects of intermittent fasting on health and performance of the body and brain, it is important to consider our evolutionary history as a species. The fundamental truth of evolution is that individuals who are best able to acquire food and avoid predation or disease are the most likely to survive long enough to pass their genes on to future generations. Some of the genes that enable success in the competition for limited food sources code for proteins that are critical for functions of the nervous system (visual or auditory acuity, learning and memory, decision making, etc.). Other genes enable sustained physical performance (running speed, strength, endurance, manual dexterity, etc.) during periods of food deprivation.

There is abundant evidence that even the most highly complex capabilities of the human brain—creativity, imagination, and language—evolved because they provided advantages in acquiring and sharing food. These capabilities include making tools for hunting, fishing, and growing food; cultivating plants and domesticating animals; and developing methods to enable the efficient distribution of food. Many if not most people think that the advanced intellectual abilities of humans are far beyond those of all other animals, but it is humbling to realize that those capabilities of the human brain have only been repurposed from what they were evolved to accomplish— survival and reproduction in environments where food was scarce and competition strong.

BRAINS AND BODIES EVOLVED TO FUNCTION BEST WHEN FOOD IS SCARCE

Success in competition for limited amounts of food has been the single most important determinant of the survival of individuals in all species and their ability to pass their genes on to a next generation. Individuals who are the most proficient in locating and acquiring food are those most likely to survive and procreate. This applies to the evolution of all animals, including *Homo sapiens*. Prominent among the capabilities of the brain that enabled our human ancestors to overcome food scarcity are the creativity that enabled them to design and manufacture tools for hunting, to control and utilize fire, and to domesticate plants and animals; the development of languages that enabled them to accumulate vast amounts of valuable information and to pass on the information across generations; and the organization of societies via governments and religions that established distribution of effort and moral standards.

As a framework for my research on intermittent fasting, I have regularly contemplated the fundamental question of how adaptations of the brain and body that enable us to perform well during periods of food deprivation provided a survival advantage during evolution. Good examples to consider in this regard are the evolution of visual acuity, speed, and agility in predatory cats and the evolution of the "social brain" in wolves that enabled their cooperation in hunting large prey. Imagine a pack of wolves surrounding a buffalo.

The notion that many of the mental and physical capabilities of humans evolved as adaptions to food scarcity is supported by considerable evidence. The oldest-known tools invented by humans were used for hunting animals for meat. And there is a vast scientific literature on foraging behaviors in herbivores and how their brains enable them to make decisions regarding when to leave their current location and seek more abundant food sources elsewhere. Such decisions are based, in part, on the energy (calorie) content of the plant sources. For example, nuts and fruits are more energy dense compared to grasses and leaves.

And then there is the fact that plants have evolved mechanisms that protect them from being consumed. Plants have a remarkable ability to produce myriad chemicals that protect their vital parts from being consumed by insects and other animals, including humans. I have come to the conclusion that the reason that vegetables and fruits, coffee, tea, and dark chocolate are good for health is that they contain bitter-tasting noxious chemicals that induce mild beneficial stress responses in our cells. For example, caffeine, curcumin (in turmeric), sulforaphane (in broccoli), and resveratrol (in red grapes) are bitter-tasting chemicals produced by the plants for the purpose of protecting them from being eaten. It was advantageous for us to be able to tolerate these "phytochemicals" so that we could acquire the energy and other nutrients in the plants.

Accordingly, we evolved multiple mechanisms that enable us to consume considerable amounts of these plants' parts and to respond positively to their bitter-tasting chemicals. I elaborate on such health-enhancing phytochemicals in chapter 6.

The cellular architecture of the human brain as we know it today was "sculpted" by evolutionary processes over millions of years of evolution. And that brain sculpting was based, at least in part, on whether a particular brain structure or type of nerve cell improved success in food acquisition or reproduction or both. Particularly important for humans are the hippocampus, which is critical for successful navigation to food sources in complex environments; the prefrontal cortex, which plays major roles in decision making and performing complex tasks; and the hypothalamus, which controls appetite, energy metabolism, and reproduction. As you read on, you will learn about neuronal circuits in these brain regions, their roles in daily life, and how they are influenced by intermittent fasting.

The ability to navigate accurately while moving through complex environments is fundamental for success in food acquisition. The hippocampus is prominent among brain regions that control navigation—knowing one's current location and remembering the paths traveled to arrive there. By recording the activity of neurons in the hippocampus of rats and mice as they move around in laboratory cages or when they are wearing virtual-reality goggles, neuroscientists have discovered something like a geographical positioning system that guides them. Specific neurons encode the animal's current position and orientation. Moreover, those neurons become active when the animal is "imagining" future navigation paths. Hippocampal neurons called "place cells" encode the animal's current location, and neurons in the entorhinal cortex (a brain region directly connected to the hippocampus) called "grid cells" are critical for planning where the animal will go next. Sequential firing of hippocampal place cells "replay" previous routes to a food-reward goal. Both forward replays and reverse replays occur, which are likely critical for the animal's ability to mentally visualize routes it has traveled so that it can easily return to a previous location. Knowing where food is most likely to be found and how to get there is fundamental for the survival of animals and humans.

The size of the prefrontal cortex is much bigger in apes and particularly in humans than in other mammals such as dogs and cats. There is convincing evidence that the expansion of the prefrontal cortex during primate evolution enabled apes that lived in the arboreal canopy in tropical forests to make critical decisions that increased their efficiency in foraging for fruits and nuts. The tropical forest comprises a wide variety of tree species that produce fruits and nuts, and those trees are typically located in different patches, where they become ripe and optimally nutritious at different times of year. These sporadically distributed food sources and their intermittent availability made decisions as to when to leave a largely depleted food patch and search for a different food patch critical for survival. Studies have shown that monkeys and human hunter-gatherers will leave a patch when its food density is reduced to the average food density for the overall area of their foraging territory. Electrophysiological recordings from neurons in monkeys' brains have revealed how neurons in different brain regions mediate foraging decisions. The studies are revealing how information is integrated between the visual cortex, prefrontal cortex, hippocampus, and motor cortex in ways that maximize the efficiency of food acquisition. In particular, strong connections between the hippocampus and prefrontal cortex coordinate spatial navigation (hippocampus) with executive decision making (prefrontal cortex).

Evolutionary changes in the musculoskeletal and physiological features of humans enabled them to walk and run long distances efficiently. As humans' physical territory expanded and the size of their social groups increased, their brains evolved advanced capabilities that enhanced food acquisition and year-round security. These capabilities include creativity, imagination, social cooperation, and language. The "food-centric" origin of these higher cognitive functions is front and center in the archaeological record. Most of the first tools manufactured by our hunter-gatherer human ancestors over a period of several hundred thousand years were for the purposes of acquiring or processing food—flaked stones, spears, bows and arrows, fire, pots, and eating utensils. And then came animal domestication, agriculture, food transportation, storage and distribution, tractors, combines, and the processed-food industry.

Interestingly, it turns out that some species of birds, including crows and parrots, also evolved advanced cognitive capabilities for food acquisition that are similar to those in humans. The word *caching* describes the behavior of hiding small collections of food (or other valuable items) in places where they are not readily found by others. By studying the caching behavior of scrub jays, which are in the corvid family, Nicola Clayton at Cambridge University has revealed these birds' remarkable humanlike cognitive abilities. The birds hide small collections of food under leaves, logs, and other such places and then return to consume the food over time. Clayton shows that these birds can remember the past, plan for the future, and interpret the behaviors of other birds. It turns out that the number of neurons in the forebrain of corvids and parrots are greater than the number in nonhuman primates with much larger brains, which may contribute to the birds' advanced intelligence. Also similar to primates, corvids have a prominent hippocampus that can generate "cognitive maps" of their environment. Their hippocampus is connected with brain structures involved in executive functions (goal-oriented planning, strategizing, and self-monitoring) and decision making. Quite notable is how birds in the crow family can understand that other individuals have mental states and intents similar to their own—an ability called "theory of mind." Moreover, they routinely use mental time travel, problem solving, and social cognition to acquire, hide, and recover foods from their caches. These birds remember what happened, when, and where based on a single past experience, which enables them to discriminate between many similar previous episodes. They even learn the "shelf life" of foods and eat the perishable foods first.

Humans are unusual in that the natural situation of food scarcity that drove the evolution of their brains changed drastically as a result of the Agricultural Revolution, which began approximately 10,000 years ago. Since then, many human societies throughout the world have experienced the luxury of food abundance. Prior to the Agricultural Revolution, the modern-day eating pattern of breakfast, lunch, and dinner was not possible. People did not wake up and have food waiting for them. They had to work for their food and, in fact, spent most of their waking hours doing so.

As the efficiency of agricultural practices and food distribution increased, more and more people engaged in new occupations unrelated to acquiring food. Organized educational systems and institutions of higher learning proliferated. Importantly, humans developed methods for interrogating nature in ways that led to a true understanding of our place in the universe and of life on earth. Science and technology flourished.

Studies of inanimate objects (physics) and of living things (biology) were launched in earnest during the Scientific Revolution in the 1500s and the Enlightenment in the late 1600s and early 1700s. Nicholas Copernicus provided evidence that the earth revolves around the sun, which was confirmed and extended by Johannes Kepler and Galileo Galilei, who described the laws of planetary motion. Then came Isaac Newton, who developed fundamental mathematical laws of physics, including gravity, inertia, friction, and so on. Religious leaders, government officials, and much of the general public scorned many of these early scientists because the scientists' discoveries contradicted information in religious texts. But most of the early scientists were religious, too, and they ultimately had to decide to believe their eyes rather than the myths of their religion. The revelations of science established what is real and what is imaginary, and they continue to do so.

In the mid-1800s, Charles Darwin and Alfred Wallace accumulated vast amounts of evidence that strongly suggested that species evolved over long periods by a process of selection for traits that provided advantages for survival and reproduction in the species' territorial niches. The evolutionary origins and principles of life on earth were revealed. During the ensuing 150 years, exponential advances in genetics, molecular biology, biochemistry, and cell biology established the mechanisms by which evolution occurs. The specific sequences of the four DNA bases (A, C, T, G) in the nucleus of our cells determines what complement of proteins our cells can produce. Proteins comprise chains of amino acids, and the sequence of amino acids in each protein is determined by the DNA sequence of the gene that encodes that protein. Each amino acid is encoded by a three-base sequence in the DNA. We inherit half of our genes from our father and half from our mother, and so each of us has a mixture of traits of both parents.

Our complement of genes and the proteins they code for determines not only our physical characteristics but also our behavioral capabilities and tendencies as well as whether we are likely to develop a particular disease. For example, if both of your parents developed heart disease at an early age, then you are also much more likely to develop heart disease compared to your friend whose parents have no heart disease. This also applies to brain disorders, including depression, anxiety, Parkinson's disease, and Alzheimer's disease. But except in rare cases of genetic mutations that cause a specific disease, the genes you inherit affect only your risk of *developing* a disease. This is good news because it provides an opportunity to modify your diet and lifestyle so as to counteract any potentially "bad" genes that you have inherited from your parents. As you continue reading, you will learn that intermittent fasting is a lifestyle modification that may protect against many of the major diseases from which people suffer—heart disease, diabetes, stroke, cancers, and a range of brain diseases.

A major advance in understanding the brain came in the late 1800s when Italian pathologist Camilo Golgi developed a method for staining nervous system tissues that resulted in only a small percentage of the neurons staining intensely black. This differentiation enabled the visualization of the entirety of the dendrites and axon of the neurons on a white background of unstained cells. Prior to development of the "Golgi staining method," it had not been established that the brain comprises billions of individual neurons connected by synapses. The Spanish neuroscientist Santiago Ramón y Cajal then used the Golgi staining method to painstakingly look at the nervous systems of many different species of mammals under a microscope and to draw the stained neurons that he observed. His research provided an unprecedented view of the complex organization of nerve cell networks in the brain and laid the foundation for subsequent elucidation of the functions of different brain regions and how neurons communicate with each other via electrical activity and the actions of neurotransmitters at synapses. Examination of Cajal's drawings of nerve cell networks in different brain regions of various animals—mice, cats, dogs, monkeys, and humans—revealed that the "cellular architecture" of these animals' brains was very similar. Thus, the main difference between the human brain and the brains of lower animals is that there are more neurons and synapses in the human brain.

Studies of the molecular neurobiology of the brain during the past 50 years have demonstrated that essentially all of the genes in neurons in the human brain are also present in the neurons of the brains of the most commonly studied laboratory animals—mice and rats. Research in my laboratory and other laboratories has identified many genes common to rodents and humans that respond to intermittent fasting in ways that improve the function and durability of cells and the organs in which the cells reside. Some of these genes encode proteins that increase the number of mitochondria (where respiration and energy-production processes occur) in cells, while others encode proteins that enhance cells' ability to repair damaged DNA, squelch free radicals, or recover from stress. These genes are relatively dormant in the cells of individuals who live over-indulgent and sedentary lifestyles. I suspect that this explains at least in part why such lifestyles increase the risk for so many diseases, including those that affect the brain.

A remarkable and somewhat troubling aspect of the prominent force of food scarcity in driving brain evolution has recently emerged from studies of the brains of domesticated animals and humans. The intelligent design of dog breeds by humans was first based on selectively breeding only the friendli-est dogs and then on selective breeding for specific body features (e.g., small or large, black or red hair, cute or rugged looking) or behavioral traits (e.g., sheep herding or fox hunting). Interest-ingly, however, if you have a pet dog, its brain is smaller than the brain of the wolves from which it evolved during the pro-cess of domestication (figure 1.1). Even if your dog's body size is as big as a wolf, its brain size is smaller. This is apparently also true of domesticated farm animals, whose brains are smaller than the wild species from which they originated. Therefore, the reduced brain size of domesticated animals has occurred regardless of the traits that we humans selected for and must therefore be the result of something about the environment that is common to all domesticated animals. I would argue that the reduction of brain size upon domestication is almost surely the result of the fact that the animals are provided food

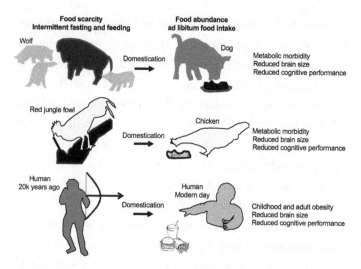

Figure 1.1

Food overabundance—which results in not having to use one's brain to acquire food—is believed to be the reason that brain size of domesticated animals and modern-day humans (since the Agricultural Revolution) has decreased compared to the brain size of their predecessors that lived in food-sparse environments.

continuously. They therefore do not experience intermittent food deprivation and so, in contrast to their counterparts in the wild, do not have to expend the mental and physical effort to acquire food. Apparently, during a relatively short number of generations, some of the neurons and synapses in brain regions that had evolved under the pressure of food scarcity are no longer necessary for survival in an environment of constant provision of food by humans.

The phenomenon of reduction in brain size during domestication may also be occurring in humans by a process of "self-domestication." By measuring the skulls of people who died more than 10,000 years ago and comparing their cranial volumes (the space that their brain occupied) to those of people living in modern societies, scientists have concluded that there has been an approximately 10 percent reduction in overall brain size in humans. It may be no coincidence that this reduction in brain size occurred relatively rapidly subsequent to the advent of agriculture and the development of effort-sparing technologies. I suspect that the disuse of brain circuits critical for success in hunting and foraging has resulted in this reduction in brain size—a "use it or lose it" scenario. Although being fluent in both spoken and written language is necessary for most occupations in modern societies, being adept at critical decision making while moving through complex physical environments is not. Because the brains of humans who lived prior to the Agricultural Revolution do not exist, we cannot compare the brains themselves, only the skulls. Nevertheless, it might be expected that brain regions previously used extensively on a day-to-day basis for hunting and foraging (motor cortex, temporal and frontal lobes) have diminished, whereas those regions used for language and abstract thought have increased in size. Several hunter-gatherer societies are still living in Africa, and it would be of considerable interest to acquire and analyze images from computerized tomography (CT) or magnetic resonance imaging (MRI) scans of their brains and to compare them to the scans of brains of people in modern (non-hunter-gatherer) societies.

The potential adverse consequences of overindulgent sedentary lifestyles on the brains of current and future generations merit serious concern. Everyone reading this book knows that being chronically overweight and sedentary does not bode well for their cardiovascular system. People with a body mass index

(BMI) greater than 25 are at increased risk for heart disease, stroke, and diabetes. This is even more of a concern for people with obesity (BMI of 30 or more). In fact, obesity is now considered a disease that requires intensive treatment. During the past 10 years, a compelling body of evidence has accrued that shows that the brains of individuals who chronically overeat do not function as well (on average) as those who moderate their food intake. Adults with obesity perform relatively poorly on various cognitive tests compared to normal-weight adults of the same age. The poorer cognitive function of people with obesity is associated with a reduction in the size of their hippocampus. Diabetes can accelerate brain shrinkage and cognitive impairment, and it is now recognized that individuals who are obese or have diabetes or both are at increased risk for Alzheimer's disease. This means that because the obesity epidemic has occurred only recently (during the past 40 years), there will be a corresponding increase in the number of individuals with Alzheimer's disease as those individuals reach their sixth, seventh, and eighth decades of life. Ironically, the Alzheimer's disease epidemic has been compounded by advances in the early diagnosis and treatment of cardiovascular disease and many cancers, which enables many people who would previously have died from these diseases when they were in their fifties or sixties to live into their seventies and eighties. Unfortunately, there are no treatments for Alzheimer's disease that even slow down its unrelenting havoc on the brain, let alone a cure. Read on, and you will learn how intermittent fasting may reduce your risk for Alzheimer's disease.

Even more disturbing than the adverse effects of chronic food overconsumption on the brains of adults are studies showing that young children and adolescents who are obese exhibit impaired learning and memory capabilities and poorer academic achievement compared to their normal-weight classmates. In America, the states with the highest prevalence of childhood obesity also have the lowest percentages of high school and college graduates. Moreover, maternal and paternal obesity predisposes offspring to poorer cognitive outcomes by "epigenetic" molecular changes—not changes in the DNA sequence of genes but rather changes in the amount and location of molecular "tags" on the DNA that can alter how much the gene is turned on or off, thereby increasing or decreasing the production of the specific protein encoded by the gene. In chapter 7, I discuss the role of epigenetics in the increased risk for autism in children born to mothers with obesity or diabetes.

But our evolutionary history offers us and future generations several reasons for optimism. It is clear that many brain structures and nerve cell networks that evolved to enable success in food acquisition can also serve in other problem-solving tasks. For example, engaging in intellectually challenging endeavors on a daily basis "exercises" neuronal circuits in the hippocampus and prefrontal cortex so that the numbers of synapses between neurons in those circuits are maintained or increased. Keeping one's brain intellectually and socially engaged may also promote the production of new nerve cells from stem cells in the hippocampus in a process called "neurogenesis." There is evidence that people in occupations that require new intellectual challenges every day are less likely to develop dementia as they grow older.

Brains have evolved to function well or even optimally in a food-deprived (fasted) state. Food overindulgence adversely affects the brain because the signaling pathways that evolved to enhance brain function under conditions of food scarcity are relatively disengaged when food is consumed throughout one's waking hours. Being overweight and sedentary can even counteract the beneficial effects of intellectual challenges on brain function and resilience. Intellectual challenges are not sufficient to achieve maximum cognitive capabilities or to sustain cognitive performance as one grows older. Research suggests that physical exercise and intermittent fasting accentuate the beneficial effects of intellectually challenging tasks by stimulating the formation of new synapses and promoting neurogenesis as much as or more than intellectually enriched environments do on their own. Therefore, although the current excessive growth trajectories of obesity, diabetes, and cognitive disabilities do not bode well for the future of human brain evolution, there is considerable room for reversing this trend.

And so as you read on, keep in mind visions of a pack of wolves surrounding a buffalo or our hunter-gatherer ancestors making spears because those images capture the essence of how living in food-sparse environments "sculpted" our bodies and brains by the process of natural selection.

FOOD-CENTRIC ORIGINS OF HUMAN CULTURE

Your own creativity within and outside of your main occupation has its evolutionary origins in your ancestor's struggles to acquire and process food. The increase in the brain size of hominids (great apes as well as extinct and current human species) coincided with increased size and complexity of social groups. Complex social behaviors that enhanced survival and reproductive success for chimpanzees and early hominids evolved in settings of foraging or hunting. Selection for the "social brain" in primates was driven in part by cooperation in food acquisition and the transgenerational transfer of skills related to food acquisition (e.g., use of sticks for termite extraction by chimpanzees, the manufacture of flaked stones by early hominids). Michael Platt's research suggests that the major expansion of the human brain coincided with new behaviors that enabled acquisition of high-calorie diets: hunting in groups, inventing weapons, and using fire to cook meat.

The term *creativity* is commonly used to refer to individuals' ability to develop new advances in particular areas of endeavor—music, art, writing, architecture, science, and so on. However, until recently in human evolution, such realms of creativity did not exist. Creativity involves not only the mental manipulation of patterns of images and sounds in ways that reveal new possibilities that can then be tested, often by trial and error, but also the use of probabilistic approaches. For example, it is likely that at some time in human evolution a creative individual imagined that affixing a small, pointed cutting stone to the end of a straight stick would increase the effectiveness of a spear in killing a large animal. Another food-acquisition-driven example is from a time when carts with wheels were invented and cattle and horses were domesticated. To ease the physical burden on the farmers harvesting crops, "tack" was invented to enable the pulling of the cart by the cow or horse.

The importance of food in the organization, function, and belief systems of the large societies into which we were born is profound. Prior to the domestication of plants and animals and to the advent of farming, humans existed in relatively small groups, in which the daily activities of all members focused on either hunting or foraging for edible plants. These activities were *everyone's* occupations. As agriculture took root (pun intended) and expanded, fewer and fewer people were required to produce the amounts of food sufficient to supply everyone in the population. Accordingly, more people had time available to devote to new endeavors not directly related to food, and new specialized occupations arose—inventing, engineering, building, garment making, doctoring, teaching, researching, and many others. Therefore, the neuronal-network-based mechanisms of human creativity that evolved to cope with food scarcity were sufficient to enable creative thinking in realms unrelated to food acquisition and processing.

Whereas nonhuman primates learn specific foraging and hunting strategies and toolmaking by mimicking their parents or other skilled adults, humans evolved language. Written and spoken languages are essential for most human occupations because they provide an efficient means for the codification and dissemination of knowledge. The communication of skills for toolmaking and hunting was likely among the factors that drove the evolution of language. Individuals with the ability to effectively communicate to a child the precise locations of food sources, methods for acquiring the food, hazards, and other salient features of their environment would increase the likelihood that the child would survive and reproduce. The ability to transfer information rapidly and with a high level of precision (e.g., "The antelope is behind the large tree" or "The rival tribe is preparing to attack our city from the north") among individuals within a family or community provided a clear adaptive advantage. Transfer of information across generations was dramatically facilitated by spoken language and writing.

HISTORICAL PERSPECTIVE ON FASTING
FOR (BRAIN) HEALTH

From early on, all of the major religions appreciated benefits of fasting for the brain and incorporated it into their ritualistic behaviors. For example, it is written in the Bible, "Is not this the fast that I choose: to loose the bonds of wickedness, to undo the straps of the yoke, to let the oppressed go free, and to break every yoke?" (Isaiah 58:6). The Qur'an states that fasting was prescribed for those who came before the followers of Mohammad (i.e., the Jews and Christians) and that by fasting a Muslim gains *taqwa*, "Godconsciousness" or "Godwariness." Fasting is believed to help promote chastity and humility, prevent sin, and prevent the outburst of uncontrolled lusts and desires and farfetched hopes. To Muslims, fasting acts as a shield with which the Muslim protects himself or herself from *jahannam* (hell). In Jainism, intermittent 36-hour fasts are thought to remove karma from the soul and to improve well-being. But there is also a history of intermittent fasting for the improvement of health and mental clarity beyond such religious traditions.

Throughout recorded history, people have noticed that health can be improved by skimping on food intake. Early medical practitioners were keenly aware of the fact that excessive energy intake is a precursor to many illnesses.

> Humans live on one-quarter of what they eat; on the other three-quarters lives their doctor.
> —Egyptian pyramid inscription, 3800 BCE

> Fasting is the greatest remedy—the physician within.
> —Philippus Paracelsus, one of the three fathers of Western medicine, circa 1525

> A little starvation can really do more for the average sick man than can the best medicines and the best doctors.
>
> —Mark Twain, circa 1890

Edward Dewey graduated from the University of Michigan Medical School in 1864, and after being a surgeon in the US Army, he started a private practice in Meadville, Pennsylvania (figure 1.2).

Dewey published the book *The No-Breakfast Plan and the Fasting Cure* in 1900, where he suggested that a healthy eating pattern is to forgo breakfast and eat only two meals each day. Dewey is to my knowledge the first proponent of 16- to 18-hour fasts every day. As it happens, it would be more than a century before data from randomized controlled trials in laboratory animals and human subjects would reveal the

Edward Dewey   Linda Hazzard   Otto Buchinger

Figure 1.2

Historical proponents of fasting. Edward Dewey espoused the benefits of short daily fasts, what is now referred to as "daily time-restricted feeding," and Linda Hazzard and Otto Buchinger established clinics for long-term fasting. Buchinger carefully monitored the health of his patients, but Hazzard did not.

health benefits of such daily time-restricted eating. Many of the conclusions Dewey drew from his careful observations of his patients and their implications for healthy lifestyles and the practice of medicine have proven correct based on recent controlled studies in animals and human subjects. Here I quote two relevant passages from Dewey's book:

> Now the American breakfast, in point of shear necessity, is believed to be the most important meal of the day, as the means for strength that is to be called out for the forenoon of labor, and believed with a force of insistence that warrants a conclusion that a night of sleep is more exhausting to all the powers than the day of labor.
>
> The no-breakfast plan with me proved a matter of life unto life. With my morning coffee there were forenoons of the highest physical energy, the clearest condition of mind, and the acutest sense of everything enjoyable.

> Not being able to give my patients clearly defined reasons for the general and local improvements resulting from a forenoon fast as a method in hygiene, it had to be spread from relieved persons to suffering friends; and according to the need, the sufferers from various ailings would be willing to try anything new where efforts through the family physician or patent medicines had completely failed; it was spread as if by contagion, among the failures of the medical profession.

My own intermittent fasting eating pattern is the one suggested by Dewey. I simply skip breakfast and eat all of my food within a 6-hour time window every day. My experiences with this time-restricted eating are pretty much the same as Dewey's, including increased ability to concentrate and be productive, particularly in the morning hours.

Linda Hazzard, who was a "disciple" of Dewey, gave fasting a bad name, however. Rather than encouraging the daily time-restricted eating pattern, Hazzard was a proponent of long fasts of many weeks or even months. Although she did not have a medical degree, she was licensed to practice medicine in the state of Washington because she was a "practitioner of alternative medicine." In the late nineteenth century, she opened a sanitarium near Seattle, where she subjected patients with a range of maladies to long fasts. She was clearly not qualified to treat patients in this manner: at least 40 of her patients died under her care, most likely of malnutrition and chronic dehydration. This outcome was well known by people living in the area, who called her sanitarium "Starvation Heights." In 1908, Hazzard published the book *Fasting for the Cure of Disease* and claimed that all of the information and evidence in it was based on solid science. It was not. Hazzard was convicted of manslaughter in 1912 and spent two years imprisoned in the Washington State Penitentiary in Walla Walla. She then moved to New Zealand, where she again treated people with fasting until it was discovered that she did not have a medical degree. She returned to Washington State, where she opened a "school of health." She died at the age of 71 during a long fast.

Upton Sinclair was a prominent American writer of both fiction and nonfiction books who received the Pulitzer Prize for Fiction in 1943. His most famous books were *The Jungle* (1904), which documented horrible, unsafe working conditions in the meatpacking industry, and *The Brass Check* (1919), which exposed and criticized "yellow journalism." The "yellow press" used exaggeration, sensationalism, and conspiracy theories to attract and rile readers. Of course, today yellow journalism is rampant on cable news networks and the internet.

But a lesser-known book written by Sinclair is *The Fasting Cure* published in 1911. It recounts his own experience with water-only fasts of several days to a few weeks' duration, and his correspondence with more than 200 other people who fasted as a treatment for a wide range of disorders, including obesity, inflammatory intestinal disorders, rheumatism, asthma, and liver and kidney disorders. For reasons that are unclear to me, at the turn of the twentieth century the emphasis was on long, infrequent fasts. However, as I describe in some detail in the remainder of this book, we now know that much shorter and more frequent fasts accrue many health benefits and that such eating patterns can be continued indefinitely. Nevertheless, several of Sinclair's conclusions ring true today. For instance, in *The Fasting Cure* he writes, "The reader will find in the course of the book a tabulation of the results of 277 cases of fasting. In this number of desperate cases, there were only about half a dozen definite and unexplained failures reported. Surely it cannot be that medical men and scientists will continue for much longer to close their eyes to facts of such vital significance as these."

Unfortunately, they did continue much longer. Until my colleagues and I began our studies of intermittent fasting in rats and mice in the 1990s, many scientists were unaware of the potentially profound effects of fasting on health and disease processes. Medical men and women have only begun to open their eyes within the past 10 years, prodded by the publication of the results of trials of intermittent fasting in human subjects and by the attention these trials have received in the media and social networks.

> There are many such curious things, about which you may read in the books of the yogis and the theosophists—who were fasting in previous incarnations when you and I were swinging about in the tree-tops by our tails.
>
> —Upton Sinclair, *The Fasting Cure*

> No phase of the experience [of fasting] surprised me more than the activity of my mind: I read and wrote more than I had dared to do for years before.
>
> —Upton Sinclair, *The Fasting Cure*

Indeed, as I elaborate later, the positive impact of intermittent fasting on concentration and cognition have been established in scientific studies.

At the time Sinclair, Hazzard, and Dewey wrote their books, more than 100 years ago, no actual data were available from controlled studies of fasting in animals or humans. All of the advice that they and other fasting enthusiasts proffered was based on personal experience and conjecture. As the titles of their books emphasize, their claims went well beyond their anecdotal observations—these writers were fasting extremists. They often jumped to unjustified conclusions based on scant or no scientific evidence but rather on their own anecdotal experiences and post hoc speculations. Nevertheless, several of their contentions have proven to be valid at least to some degree. For example, Dewey attributed many diseases to excessive eating, and he encouraged daily fasts of 16–18 hours.

Otto Buchinger was a German physician who is well known in western Europe for his research in the early twentieth century on fasting for the treatment of diseases (figure 1.2). Although he apparently documented many of his case studies, he did not publish them and did not perform any randomized controlled trials. He founded several fasting sanatoriums, including one in Bad Pyrmont and one in Überlingen, Germany, where people with a range of maladies would go once a year to experience supervised long fasts of 10 to 14 days. These fasting spas are still widely respected and have outstanding safety records. Françoise Wilhelmi de Toledo, who is the director of the Überlingen fasting center, has recently published findings showing that during a 10-day fast people exhibit reductions in blood glucose, insulin, hemoglobin A1c, total cholesterol, and triglyceride levels. However, these beneficial changes during the fasting period may not persist when the people return to their normal daily eating routines. Emerging evidence described in the remainder of this book shows how much shorter and frequent fasting periods—an 18-hour fasting period every day, for example—can improve health in a sustainable manner.

2 INTERMITTENT FASTING SLOWS AGING

> It is truly amazing that a complex organism, formed through an extraordinarily intricate process of morphogenesis, should be unable to perform the much simpler task of merely maintaining what already exists.
>
> —Francois Jacob, *The Possible and the Actual*

All animals age and die. As they age, the structures and functions of cells and organs progressively deteriorate. This deterioration accelerates rapidly after an individual reaches the end of his or her reproductive period. This makes sense because evolution selects for traits that enable survival and reproduction. From the perspective of survival of the species, keeping old individuals alive indefinitely would be detrimental. The old individuals would consume part of the limited amounts of food available for children and adults of reproductive ages. Several major changes that occur in brain cells during aging have been documented and are shown in figure 2.1. At the center of these changes is the dysfunction of mitochondria (where respiration and energy-production processes occur) and a consequent cellular adenosine triphosphate (ATP) energy deficit.

Figure 2.1

The "hallmarks of brain aging," with dysregulation of cellular energy metabolism shown in the ring around the center. The author's research has provided evidence that intermittent fasting can counteract these hallmarks of aging. A modified version of this illustration by the author was published in M. P. Mattson and T. V. Arumugam, "Hallmarks of Brain Aging: Adaptive and Pathological Modification by Metabolic States," *Cell Metabolism* 27 (2019): 1176–1199.

The age-related decline in mitochondrial function is exacerbated by increased damage to molecules by free radicals and the cells' impaired ability to repair the damage or remove the damaged molecules. The cells' ability to respond adaptively to stress is impaired during aging. These adversities of cellular aging in the brain have several consequences, including neuronal networks' compromised ability to regulate their electrical activity properly, cell senescence, inflammation, and depletion of stem cells. This chapter describes how intermittent fasting can help cells forestall the aging process.

The term *intermittent fasting* can be traced to an article published in the *Journal of Nutrition* in 1946 by Anton Carlson and Frederick Hoelzel, who were then at the University of Chicago. They provided evidence that intermittent fasting can increase the average lifespan of rats, but the study was problematic in that Carlson and Hoelzel did not perform statistical analysis of their data. The first clear evidence that intermittent fasting can have major effects on health and longevity came from a study by Charles Goodrick, Don Ingram, and Nancy Cider at the National Institute on Aging in Baltimore in the early 1980s. They found that the lifespan of rats maintained on an every-other-day fasting regimen was increased by more than 80 percent compared to rats fed ad libitum (with food available to them all the time). Eight years later they published the results of another study showing that intermittent fasting also increases lifespan in mice. In the early 1990s, I became aware of that research at the National Institute on Aging as well as of the work by Rick Weindruch at the University of Wisconsin and by others who reported that daily calorie restriction can also increase the lifespan of rats and mice. It was clear that compared to animals that have food available to them 24/7, animals given a daily allocation of an amount of food less than what the they would normally consume live longer. The amount of lifespan extension is roughly proportional to how long the animals are on the food-restricted diet. That is to say, lifespan extension is greater when the daily or every-other-day food deprivation is begun when the animals are young adults, less when started in middle age, and much less or negligible when started in old age.

During the past 30 years, researchers at the National Institute on Aging and the University of Wisconsin have been studying the effects of daily calorie restriction on aging in monkeys. They have found that providing the monkeys with less food than they would normally eat slows the aging process and the burden of diseases as they age. Caloric restriction can extend the monkeys' lifespan, and this extension is associated with a reduced body weight compared to monkeys whose calories are not restricted. In these studies, caloric restriction was particularly effective in counteracting the accelerated aging caused by a diet with a high sugar content.

Luigi Fontana is at the forefront of research on how calorie restriction affects human physiology and disease risk. Some of his studies involve randomized controlled trials of daily calorie restriction, and he also studies a group of people who have been intentionally keeping their daily calorie intake very low and eating a variety of health foods for many years, a group that he has dubbed CRONIES, for "calorie restriction with optimal nutrition." Their BMIs are between 17 and 19. Although blood ketones have not until recently been measured in such studies, the calorie-restricted people are clearly using fats as energy because their abdominal fat levels are decreased. Fontana's studies of the CRONIES have shown that daily food restriction improves multiple indicators of cardiovascular disease risk, including reductions in low-density lipoprotein (LDL) cholesterol levels, increased high-density lipoprotein (HDL) cholesterol levels, and improved blood sugar regulation.

The obvious question posed by the "anti-aging" effect of intermittent fasting is: How does it work? The answer that has emerged during the past twenty years is that intermittent fasting causes complex and coordinated changes in cells that counteract aging. It has also become clear that the converse is true—excessive energy intake (i.e., overconsumption of food, especially those high in sugar and fat) accelerates the aging process by disabling the same cellular processes that are activated by intermittent fasting.

IMPROVING GLUCOSE AND FAT METABOLISM

Physicians use the term *metabolic syndrome* to describe an increasingly common unhealthy clinical profile. Someone with metabolic syndrome has excessive amounts of abdominal fat, insulin resistance, elevated cholesterol and triglyceride levels, and elevated blood pressure.

When a person eats a meal, her blood glucose levels increase, and cells in the pancreas that produce insulin sense the increased glucose levels and release insulin into the blood. In a healthy person, the insulin stimulates cells in the liver, muscles, and other organs to rapidly take up glucose, the blood glucose level returns to the low-normal levels, and then insulin levels are also reduced. In a person with insulin-resistant cells, however, the liver, muscles, and other organs become unresponsive to insulin, and so blood glucose and insulin levels remain elevated even when the person has not eaten recently. Your doctor can determine whether you have insulin resistance by measuring levels of glucose and insulin in blood drawn after an overnight fast. She or he will multiply the insulin concentration and the glucose concentration to determine the homeostatic model assessment of insulin resistance (HOMA-IR). If your HOMA-IR is higher than 1.9, you have insulin resistance.

The major cause of metabolic syndrome is an overindulgent, sedentary lifestyle that is devoid of the periods of the negative energy balance that occurs during fasting and exercise. In other words, metabolic syndrome occurs in people with lifestyles that result in calorie intake exceeding calorie expenditure on a daily basis. They rarely or never fast for a sufficient period to trigger the metabolic switch from glucose to ketones. The good news is that metabolic syndrome can be completely reversed by the adoption of a lifestyle that includes intermittent fasting and exercise. Studies of animals and humans have shown that intermittent fasting alone can reverse metabolic syndrome. In studies of mice by Mike Anson and Zhihong Guo at the National Institute on Aging in 2003, it was discovered that intermittent fasting improves glucose regulation. They found that when mice are maintained on an every-other-day fasting regimen for several months, their blood glucose and insulin levels are greatly reduced compared to the levels in mice that had food available to them 24/7. In that study, the mice did not lose weight on intermittent fasting because on the days they were fed ad libitum, they consumed almost twice as much food as they normally would. This was the first evidence that health benefits of intermittent fasting can occur even without an overall reduction in calorie intake.

The risk of developing insulin resistance and metabolic syndrome increases during aging. Moreover, experiments have shown that feeding rats or mice a diet that causes insulin resistance, such as one with high amounts of sugars and saturated fats, accelerates the aging process. The insulin-resistant animals exhibit signs of aging, including motor and cognitive impairment, tissue inflammation, and the accumulation of molecules damaged by free radicals in their cells. Age-related diseases such as cancers and kidney failure occur at an earlier age in animals fed diets that cause insulin resistance. One consequence of insulin resistance and elevated glucose levels is that sugar molecules bind to many proteins in a process called "glycation." In fact, a common way doctors diagnose and monitor patients with prediabetes and diabetes is to measure levels of glycated hemoglobin, a test commonly referred to "Hemoglobin A1c" or simply "A1c." Higher amounts of glucose in the blood will result in higher amounts of glucose bound to hemoglobin. A1c levels provide a measure of the average blood glucose levels over a three-month period. A1c levels are also an indirect indicator of aging and of risk for an earlier onset of age-related diseases. Glycation of proteins is accelerated by free radical production, which causes sugar to attach to proteins. In turn, glycated proteins can cause oxidative stress and inflammation, thereby creating a vicious cycle that accelerates aging. Multiple studies have shown that intermittent fasting can reduce A1c levels in humans.

Courtney Peterson, Eric Ravussin, and their colleagues at the Pennington Biomedical Research Center in Louisiana performed a controlled trial of intermittent fasting in men with prediabetes. Each day participants in the intermittent fasting group ate all of their food within a 6-hour time window, whereas those in the control group ate three meals during a 12-hour period. The researchers found that compared to the 12-hour eating window, a 6-hour eating window results in improvements in insulin sensitivity and blood pressure as well as a decrease in oxidative stress. Their findings suggest that daily time-restricted eating may prevent the development of diabetes in people who are at high risk.

Satchin Panda has performed many important studies of the effects of daily time-restricted eating on the health of mice. His research on intermittent fasting arose from his interest in circadian rhythms, the regular ebbs and flows of cellular metabolism and functions during the 24-hour period of one rotation of our planet. These circadian rhythms are strongly influenced by cycles of daylight and darkness, but they are also attuned to the timing of food intake. As with other diurnal animals, humans evolved consuming most or all of their food during the daytime, and sleeping during the night, whereas nocturnal animals such as racoons, rats, and mice evolved to do the opposite. Panda's research has shown that when mice are allowed to eat during only an 8-hour time period in the nighttime (their normal active period), they will not develop obesity. However, when they are provided food during an 8-hour time period in the daytime (their normal sleeping period), they do become obese. These findings are consistent with the notion that people who eat in the nighttime are more likely to develop insulin resistance compared to those who eat all their food during the daylight hours. Thus, a daily routine of early to bed and early to rise plus no food consumption after dark will keep one's circadian rhythms in an optimal state.

A question that arose from my own and Panda's studies was whether most if not all of the hundreds of studies of "calorie restriction" were also studies of intermittent fasting. The reason for thinking they might be is based on the method used to precisely reduce the daily calorie intake of laboratory animals. First, each animal's average ad libitum daily food intake is measured. Then the animals are divided into two groups. The control group continues to be fed ad libitum, whereas the other group is provided with less food than they would normally eat—typically 30 or 40 percent less. The entire daily allotment of food is provided all at one time each day. Because the rats in the calorie-restriction group are receiving much less food than usual, they consume all of the food pellets within 4–6 hours of being given the food. They therefore fast for 18–20 hours every day, and so their ketone levels are elevated.

In 2019, Sarah Mitchell and Rafa de Cabo reported that daily fasting can increase the lifespan of mice independently of calorie intake and composition of the diet. Combining calorie restriction with intermittent fasting, however, results in greater improvements in health indicators and longevity compared to intermittent fasting without calorie restriction. Nevertheless, intermittent fasting was effective even when calorie intake was not reduced. The combination of intermittent fasting and calorie restriction was very effective in lowering blood glucose and insulin levels in the mice. Further studies will be required to better understand what contributions to enhanced longevity can be made by a reduction in calories alone versus the metabolic switching and elevation of ketones that occurs with intermittent fasting.

COUNTERACTING FREE RADICALS

One of the first theories of aging was championed by Denham Harmon, who proposed in 1956 that the continuous production of free radicals that occurs in the mitochondria of cells results in progressive damage to DNA and other molecules and that this oxidative "wear and tear" is the cause of aging. More than seventy years later, the free-radical theory has proved valuable in advancing an understanding of the molecular basis of aging. However, free radicals are only a small piece of the very complex puzzle of aging (figure 2.1). Cells throughout the body and brain continuously experience damage to molecules by free radicals, and during aging the number of molecules damaged increases. The three main free radicals produced in cells are superoxide anion radical ($O_2\cdot$), hydroxyl radical ($OH\cdot$), and peroxynitrite anion radical ($ONOO^-\cdot$) (figure 2.2). The oxygen atom in each of these molecules has an unpaired electron, which is denoted by the superscript dot ($\cdot$) placed at the end of the molecular formula. Because these oxygen molecules have an unpaired electron, they try to "steal" an electron from other molecules in a process called "oxidation." It is the same process by which oxygen in the air causes the rusting of iron and the rapid browning of a freshly cut apple. Some molecules in cells are particularly susceptible to attack by free radicals, including the nucleic acid guanine in DNA, the amino acids cysteine and lysine in proteins, and the carbon-to-carbon double bonds ($C=C$) in the fats of cell membranes.

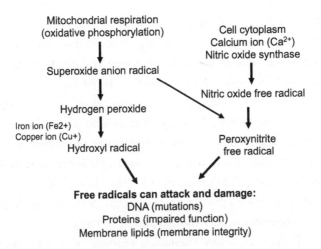

Figure 2.2

Sources of the production of damaging free radicals in cells. The major source of oxygen free radicals is the mitochondrial electron transport chain, where the superoxide anion radical is formed. Superoxide can be converted to hydrogen peroxide (which is not a free radical) by enzymes called "superoxide dismutases." In the presence of even very low amounts of free (ionic) iron or copper, hydroxyl radical is produced from hydrogen peroxide. Another source of free radicals is the enzyme nitric oxide synthase, which is activated in the cell cytoplasm by ionic calcium. Nitric oxide can interact with superoxide to produce the free radical peroxynitrite. Hydroxyl radical, nitric oxide, and peroxynitrite can attack and damage DNA, proteins, and membranes of cells. However, studies have shown that intermittent fasting can protect cells against damage by free radicals and stimulate the repair of damaged molecules.

You may be surprised that although excessive levels of free radicals can damage cells, lower levels of free radicals are actually critically important in the function of cells and the organs in which they reside. The evolution of life on earth occurred in an atmosphere with a high level of oxygen—a free-radical-rich environment. Not only did cells evolve ways of neutralizing the free radicals such as antioxidant enzymes and molecules such as glutathione, but they also put some free radicals to good use as sensors of stressful conditions and signals that initiate the cells' adaptive responses to oxidative stress. For example, the neuroscientist Eric Klann studied mice that overexpress an antioxidant enzyme called "superoxide dismutase." Surprisingly, he found that these mice have impaired learning and memory as well as dysfunction of synapses in the hippocampus. When neurons are electrically active, superoxide levels increase, and that superoxide helps the neurons respond adaptively in ways that are actually critical for the strengthening of synapses that occurs during learning and memory.

Perhaps the most fascinating example of a free radical serving important functions throughout the body and brain is nitric oxide, which is a free radical and a gas. In 1998, the Nobel Prize in Physiology or Medicine was awarded to three American scientists—Robert Furchgott, Ferid Murad, and Lous Ignarro—for "their discoveries concerning nitric oxide as a signaling molecule in the cardiovascular system," according to the press release issued by the Nobel Committee. The reason that nitroglycerin relieves chest pain in patients with coronary artery disease is that it causes relaxation of the smooth muscle that surrounds the artery, thereby increasing the diameter of the artery and so increasing blood flow to the heart. It turns out that it is the nitric oxide released from the nitroglycerin that causes the blood vessels to relax. Remarkably, nitric oxide is also produced in vascular endothelial cells, which are the cells that line arteries. When the ability of those cells to produce nitric oxide is inhibited, blood vessels constrict, and blood pressure goes sky high. We discovered that intermittent fasting reduces blood pressure in rats and mice, and Doug Seals at the University of Colorado in Boulder provided evidence that nitric oxide is involved in the reduction in blood pressure.

Working at Johns Hopkins University, Sol Snyder has made many important discoveries during the past 50 years, including identifying a receptor on the surface of neurons to which opioids bind. Snyder also established the functions of several gases in the brain that he has dubbed "gasotransmitters." They include nitric oxide, carbon monoxide, and hydrogen sulfide. In the late 1980s and early 1990s, David Bredt, who was then training in Snyder's laboratory, showed how nitric oxide is produced in neurons. Stimulation of neurons at synapses by the neurotransmitter glutamate results in calcium influx through channels in the membrane. Once inside the neuron, the calcium activates an enzyme called nitric oxide synthase, which produces nitric oxide.

Snyder's laboratory also discovered that the gases carbon monoxide and hydrogen sulfide are produced by brain cells and play important roles in communication between neurons. It is astonishing that our brain cells normally produce carbon monoxide, considering that high levels of carbon monoxide in inhaled air can result in asphyxiation and death. Ongoing research in several laboratories is aimed at determining whether carbon monoxide and hydrogen sulfide mediate effects of intermittent fasting on health. At the National Institute on Aging, we discovered that intermittent fasting increases the production of a protein called "heme oxygenase 1," which turns out to be the enzyme that produces carbon monoxide. Research by Jay Mitchell at Harvard University has convincingly shown that hydrogen sulfide is produced in cells in response to caloric restriction, and he also has evidence that in animal models this gas mediates the protective effect of intermittent fasting against liver injury.

The importance of free radicals in normal cell functions explains, at least in part, the fact that clinical trials of vitamins E and C in patients with a wide range of age-related diseases, including cancers, heart disease, and Alzheimer's disease, have failed. In fact, swamping cells with such antioxidants has even been shown to have adverse effects on healthy people. For example, vitamin E can block the ability of muscle cells to respond to exercise training by increasing their strength and endurance because the free radicals produced when muscles are active function as a signal that is critical for the changes in the muscle cells that result in enhanced endurance and strength.

As part of our research on brain aging, my laboratory worked to understand if and how free radicals might cause dysfunction and degeneration of neurons in Alzheimer's disease.

Working with chemist Alan Butterfield and neurologist Bill Markesbery (figure 2.3) at the University of Kentucky, I found that the amyloid β-peptide (Aβ) causes free-radical production in neurons and in this way may render the neurons vulnerable to degeneration in Alzheimer's disease. By causing oxidative stress, Aβ impairs the function of several vital proteins in the membrane of neurons and their synapses, which compromises the neurons' ability to control their excitability and to produce enough ATP to function properly. I found that high concentrations of Aβ can kill neurons and that lower concentrations, although not killing the neurons, greatly increase their vulnerability to being killed by a process called "excitotoxicity." Excitotoxicity involves excessive

Figure 2.3

Bill Markesbery (circa 1992) and an example of a brain section from an Alzheimer's patient that he stained to visualize amyloid plaques (*arrowhead*) and neurofibrillary tangles (*arrow*). Markesbery was the author's mentor and collaborator at the Sander-Brown Center on Aging at the University of Kentucky.

excitation of neurons by the neurotransmitter glutamate. For more information on glutamate and excitotoxicity, see the section on Alzheimer's disease in chapter 3 and the section on the brain in chapter 4.

Superoxide dismutase 2, or SOD2, is an antioxidant enzyme located in the mitochondria. It is particularly important in protecting cells because it rapidly removes the superoxide free radicals produced during oxidative phosphorylation, which is the process by which oxygen and glucose are used to produce ATP. SOD2 is the most important antioxidant enzyme in cells and might therefore be expected to play a key role in counteracting aging. Both intermittent fasting and running can enhance the activity of SOD2 in neurons in the brains of mice. By enhancing the ability of mitochondria to rapidly remove free radicals, SOD2 can protect neurons against dysfunction and degeneration in mouse models of Alzheimer's disease, epileptic seizures, and Huntington's disease. But intermittent fasting does much more than this to counteract the adversities of aging.

Research has shown that free radicals do play a role in aging, but because they are also important in low amounts for normal cell function and health, it is not beneficial to ingest high amounts of antioxidants such as vitamins E, A, and C that directly squelch free radicals. In fact, studies of what happens in muscle cells during and after exercise have shown that the increased production of free radicals acts as a signal that triggers multiple beneficial adaptive responses in the cells. These adaptations include the turning on of genes that actually reduce oxidative stress and stimulate mitochondrial biogenesis, a process that increases the number of healthy mitochondria in the muscle cells. In these ways, intermittent fasting and exercise can enhance the ability of cells to cope with oxidative stress and become stronger for having experienced the stress.

Interestingly, it turns out that most and perhaps all of the chemicals in vegetables, fruits, and spices that are good for health exert their beneficial effects on cells in a manner similar to that of exercise and fasting. Such "phytochemicals" in plants are produced for the purpose of keeping insects, herbivores, and omnivores such as humans from eating any or very much of the plant. Because plants can be a good source of various nutrients for animals, the cells in herbivores and omnivores evolved so that they respond to the noxious chemicals in plants with an enhanced ability to cope with stress, including oxidative stress.

I elaborate on this remarkable fact and its implications for optimizing one's diet in chapters 4 and 6.

CELLULAR REPAIR AND GARBAGE DISPOSAL

As you read this sentence, free radicals are attacking and causing damage to DNA in cells throughout your body and brain. But don't worry. Your cells have evolved elegant ways of efficiently repairing the damage. You will recall that DNA is a twisted (helical) double strand or string of only four different nucleic acid bases: adenine, thymidine, guanosine, and cytosine (A, T, G, C). The G on one strand binds only to a C on the other strand, while an A binds only to a T. For sequences of DNA that encode for protein, which are called "genes," three consecutive bases code for one amino acid of the protein. The nucleic acid bases can be and often are attacked by oxygen free radicals. If the damaged base is not removed and replaced with a new one, a mutation may occur that could eventually result in cancer or in neurons that can alter gene expression or that could even cause the neurons to die. Several DNA repair enzymes work together to quickly recognize, remove, and replace a damaged base in DNA.

Cells' reduced ability to repair DNA that has been damaged by free radicals appears to be a particularly important factor in aging. Several remarkable and rare premature-aging syndromes are caused by mutations in DNA repair proteins. Wilhelm Bohr's research has greatly advanced an understanding of how defects in cells' ability to efficiently repair DNA cause two premature-aging disorders—Cockayne syndrome and Werner syndrome. Even children with Cockayne syndrome exhibit signs of aging as their hair grays, their skin wrinkles, and their hearing and vision decline. They usually die before their twentieth birthday. The signs of aging appear somewhat later in people with Werner syndrome, usually when they are in their twenties or thirties, and most die before the age of 50. With help from me and Bill Markesbery, Bohr and colleagues showed that DNA repair is also compromised in brain cells of patients with Alzheimer's disease and mild cognitive impairment. We also found that DNA repair is impaired in neurons in mice genetically engineered to accumulate amyloid plaques and neurofibrillary tangles in their brains, both of which are present in Alzheimer's patients. Moreover, Bruce Yankner at Harvard has shown that damage to DNA increases in neurons during aging and often occurs in genes that encode proteins critical for the proper regulation of neuronal network function. Therefore, it is reasonable to consider the possibility that enhancing neurons' ability to repair damaged DNA may protect against Alzheimer's disease.

In 2015, the Nobel Prize in Chemistry was awarded to three scientists—Tomas Lindahl, Paul Modrich, and Aziz Sancar—for "having mapped, at the molecular level, how cells repair damaged DNA and safeguard the genetic information," according to the press release by the Nobel Committee. Remarkably, however, no one had asked how a cell's ability to repair DNA might be improved. Working with me and Bohr, postdoc Jenq-Lin Yang determined whether enhanced DNA repair can occur as an adaptive response to good stressors, including exercise and intermittent fasting. When cultured brain neurons were excited with the neurotransmitter glutamate, their ability to repair damaged DNA was improved. Interestingly, shortly after the neurons were stimulated with glutamate, a temporary increase in DNA damage occurred but was then rapidly repaired. What might this process mean for what happens in your brain as you read these words and think about their meaning? Because while you read and think there is increased activation (by glutamate) of neurons in your visual cortex, hippocampus, and brain regions involved in language comprehension, those neurons are subjected to a modest increase in free radicals and DNA damage. However, in response to this mild stress, the neurons enhance their ability to repair DNA damage and are actually better off for having been "exercised."

Further experiments showed that a protein called "brain-derived neurotrophic factor," or BDNF, stimulates DNA repair in cultured neurons. BDNF is a protein that is well known for its ability to stimulate the formation of synapses between neurons, and it plays important roles in learning, memory, and mood. Moreover, in mice models, running-wheel exercise, which increases BDNF production by neurons in the brain, has also been shown to enhance DNA repair in cells in the hippocampus and cerebral cortex. Because intermittent fasting also stimulates BDNF production in these neurons, it is expected that intermittent fasting can enhance the neurons' ability to repair damaged DNA. The take-home message is clear—exercise your neurons, exercise your body, and fast intermittently. With these three actions, your cells will improve their ability to protect their DNA. Enhanced DNA repair is one important anti-aging mechanism of intermittent fasting.

You are a completely different person physically than you were a year ago and mostly different than even a week ago! The proteins that form the structure and carry out the functions of your cells typically become damaged over time and are then removed and replaced with a pristine protein. Different proteins have different lifetimes that range from a few hours to many weeks. This fact is particularly interesting to a neuroscientist such as me because of its implications for learning and memory. We are able to remember specific events for decades afterward even though the proteins in the neurons of the circuits in which these memories were initially stored have been replaced with new proteins. Although we are far from knowing the exact molecular nature of memories, we do know that all cells, including nerve cells, have the ability to recognize and then destroy or recycle damaged molecules. The damaged molecules are then replaced by newly manufactured, pristine molecules. We now understand the major pathways by which damaged molecules are routed to the cellular garbage disposal and recycling centers. We also know that these garbage disposal and recycling centers often become overloaded and dysfunctional in cells during aging but also that intermittent fasting can ameliorate this problem.

In 2016, the Nobel Prize in Physiology and Medicine was awarded to the Japanese scientist Yoshinori Ohsumi for "discovering and elucidating mechanisms underlying autophagy, a fundamental process for degrading and recycling cellular components," according to the Nobel Committee press release. Cells have evolved the ability to recognize and efficiently remove damaged proteins, membranes, and even entire mitochondria. The question that Ohsumi answered was "Exactly how do cells remove damaged molecules?" The answer is by an elegant molecular garbage disposal and recycling process called "autophagy"—self (*auto*) eating (*phagy*). The disposal of dysfunctional and damaged mitochondria is called "mitophagy." Damaged molecules and mitochondria are transported to a membrane-bound organelle called the "lysosome." The inside of the lysosome is filled with acid and special digestive enzymes, very similar to what it is like inside the stomach. In both your stomach and lysosomes, proteins are broken down into amino acids, and cell membranes are separated into their individual lipid components, such as cholesterol and fatty acids. In the case of the gut, the amino acids and lipids are moved into the bloodstream and distributed to cells throughout the body. In the case of lysosomes, they can release amino acids and lipids, which are then used for the production of new proteins and membranes—a recycling process.

> Pharmacological or genetic manipulations that increase life span in model organisms often stimulate autophagy, and its inhibition compromises the longevity-promoting effects of caloric restriction.
>
> —David Rubinsztein, Guillermo Mariño,
> and Guido Kroemer, "Autophagy and Aging"

The PubMed website, which is curated by the National Institutes of Health (NIH), provides the world's largest collection of journal articles in the fields of medicine and biomedical research. Entering the search terms *autophagy* and *caloric restriction* or *intermittent fasting* retrieves nearly 700 articles. It turns out that fasting is the most powerful natural means of stimulating autophagy. This has been shown in a wide range of animals, including worms, flies, mice, and humans. Among the earliest evidence that autophagy is important in the anti-aging effect of intermittent fasting came from the laboratory of

Ettore Bergamini at the University of Pisa in Italy. Bergamini and his colleagues provided evidence that autophagy is critical for beneficial effects of caloric restriction on the liver of rats. During the fasting periods of intermittent fasting, autophagy and mitophagy are stimulated in the cells of many different tissues, including the liver, muscle, and the brain. At the same time, to conserve resources during fasting, any new proteins and membranes that are synthesized use mainly amino acids and lipids recycled from the lysosomes. Thus, during fasting cells switch from using glucose to using ketones (derived from fat cells) as a major energy source, while they simultaneously shift to a molecular recycling stress-resistance mode.

In some organs, such as the skin and the lining of the intestines, old cells are regularly removed and replaced with new ones. Because old cells that are not functioning well can be removed and replaced with new ones (which are produced from stem cells), autophagy and mitophagy are not as critical for maintaining optimal function of those organs. However, the central nervous system (brain and spinal cord) is very different because the neurons in that system's tissues must survive and continue functioning well throughout life. Therefore, the function of lysosomes and the molecular machinery of autophagy and mitophagy are absolutely vital for neurons. Recent research has provided evidence that impaired autophagy and mitophagy are fundamentally involved in the demise of neurons in Alzheimer's and Parkinson's diseases. By studying postmortem brain tissue from Alzheimer's patients (and age-matched neurologically normal controls) and animal models, neurologist Ralph Nixon has found evidence that the dysfunction of lysosomes occurs early in the disease process and likely contributes to the abnormal accumulation of amyloid plaques and neuro-fibrillary tangles in the brain. Intermittent fasting can stimulate autophagy in neurons and may in this way protect against Alzheimer's disease.

The importance of impaired autophagy in many different diseases of aging has prompted pharmaceutical companies to invest in research aimed at discovering chemicals that stimulate autophagy. One such chemical called "rapamycin" was originally isolated from samples of bacteria taken from Easter Island nearly 50 years ago and was then developed as a drug that suppresses the immune system in organ-transplantation patients. Three independent studies showed that administering low doses of rapamycin—sufficient to stimulate autophagy without adverse side effects—significantly increased the lifespan of mice. This was the first clear evidence that it is possible to slow aging with a chemical. Although I am skeptical that any drug or dietary supplement can fully mimic the anti-aging effects of intermittent fasting without significant side effects, this line of research will continue and may lead to many clinical trials in patients at risk for or currently in the grip of various age-related diseases. I describe some of the most promising chemicals in chapter 9.

REDUCED INFLAMMATION

The immune system consists of so-called innate and humoral cells that have evolved to respond to invading pathogens (bacteria, viruses, and parasites) and tissues damaged by a physical injury. Macrophages are the main cells of the innate immune system and are present in tissues throughout the body.

Macrophages constantly patrol the tissues within which they reside as they move around "looking" for invading pathogens or damaged cells. When a macrophage encounters a pathogen, it releases chemicals that damage the pathogen, and then it engulfs and "eats" the pathogen by a process called "phagocytosis." The macrophage also sends signals to other macrophages to alert them that there are pathogens or damaged cells that need to be removed. These signals include numerous proteins called "cytokines," among which tumor necrosis factor (TNF) and several "interleukins" have been the most intensively studied. In contrast to macrophages that reside within a certain territory of an organ, humoral immune cells circulate in the blood—they are the "white blood cells" or lymphocytes. Unlike cells of the innate immune system, cells of the humoral immune system have the ability to "remember" a previous infectious agent and rapidly produce antibodies against that pathogen in the event of a subsequent infection. This is how vaccines work. The person is administered a virus or bacteria that has been inactivated or a specific molecule from the pathogen, such as the "spike protein" of COVID-19. Cells of the humoral immune system generate cells that produce antibodies against the pathogen, and those antibodies cause the destruction of the pathogen.

In healthy young people, both the innate and the humoral immune systems are largely quiescent in the absence of an infection or tissue injury. However, during normal aging the innate immune system often becomes abnormally and chronically activated. This type of inflammation is called "sterile inflammation" because it occurs in the absence of an infection or a physical injury. Arthritis is a prominent example of the accumulation of macrophages and lymphocytes in affected joints. The local inflammation irritates sensory nerve endings in the joint, resulting in pain. But inflammation is also involved in other major diseases of aging, including cardiovascular disease, diabetes, Alzheimer's disease, and cancers. The narrowing of the coronary arteries that underlies heart disease results from free-radical-mediated damage to the blood vessel wall, to which macrophages respond and accumulate in that area of the vessel. The macrophages accumulate cholesterol, which builds up over years. The aging brain is also vulnerable to inflammation. In Alzheimer's disease, macrophage-like cells called "microglia" that reside in the brain become activated in the amyloid plaques, and this "neuroinflammation" is believed to contribute to the degeneration of neurons in this disease. Cancer cells often cause inflammation of the tissue where the tumor grows, and they can often evade an attack by the immune system. In fact, one of the most exciting recent advances in cancer research is the development of treatments that enable the immune system cells to recognize the cancer cells as abnormal and thus to kill them.

Contributing to the tissue inflammation that occurs during aging is the accumulation of senescent cells. A senescent cell is characterized by an inability to divide, an increase in size, the loss of normal function, and the production of proinflammatory cytokines, including TNF. Senescent cells can be identified by examining thin slices (sections) of a tissue under a microscope. During normal aging, senescent cells increase in numbers in many tissues, including that of the skin, liver, and brain. The senescent cells produce a protein called "p16," which is not present in normal cells. Darren Baker, Jan van Deursen, Jim Kirkland, and colleagues genetically engineered mice such that any cell that has p16—that is to say, any senescent cell—can be selectively killed. By studying these mice, they have elegantly documented the critical roles played by cell senescence in the normal aging of the heart, lungs, and other organs. In addition, they have published studies that implicate cell senescence in the pathogenesis (how a disease develops) of metabolic syndrome, liver disease, and osteoarthritis.

Research in my laboratory suggests that in Alzheimer's disease cell senescence occurs in a certain type of glial cell in the brain. The axons of neurons in the brain are normally wrapped by cells called "oligodendrocytes" that enable rapid electrochemical communication in neuronal networks. Oligodendrocytes can be produced from stem cells called "oligodendrocyte progenitor cells" (OPCs) that are distributed throughout the brain, where they remain dormant unless called into action to repair a damaged axon. Peisu Zhang discovered that in Alzheimer's disease OPCs move into amyloid plaques, where instead of forming new oligodendrocytes, they behave like inflammatory immune cells. The amyloid protein causes the OPCs to undergo senescence, and so they can neither divide nor become new oligodendrocytes. Instead, the rogue OPCs produce inflammatory cytokines, which are known to adversely affect the function and structural integrity of neuronal networks. Similar to what was observed in Alzheimer's patients, senescent OPCs also accumulated in amyloid plaques in the brains of mice genetically engineered to produce excessive amounts of human amyloid protein. Examination of thin sections of the brains of these mice in an electron microscope revealed that neurons degenerated when they were close to senescent OPCs.

Might it be possible to remove the senescent OPCs from the brains of the Alzheimer's mice? Using a combination of two chemicals that had been shown to remove senescent cells from the skin of old mice—dasatinib, a cancer drug, and quercetin, a chemical present in some vegetables and fruits—Zhang found that she could remove senescent OPCs from the brain, Importantly, the treatment also lessened inflammation in the brain and improved the learning and memory of the Alzheimer's mice. These findings pave the way for clinical trials of so-called "senolytic" treatments in patients with Alzheimer's. The goal of a senolytic treatment is to selectively remove the unwanted senescent cells, which then results in less inflammation and stress on the normal health cells in the tissue.

Intermittent fasting can reduce chronic pathological inflammation without compromising the immune system's ability to respond to infection or physical trauma. Studies of mice have shown that intermittent fasting reduces inflammation as well as disease pathology and symptoms in models of a range of disorders, including obesity, cancer, inflammatory bowel disease, stroke, and multiple sclerosis. It does so by reducing tissue damage, suppressing the activation of macrophages, and reducing the production of the inflammatory cytokines. In 2019, Valter Longo reported that intermittent fasting can reduce intestinal inflammation in a mouse model of inflammatory bowel disease, and this reduction was associated with a shift in the gut microbiota to more healthy types of bacteria. Remarkably, he found that transplantation of feces from mice on a time-restricted feeding regimen reduced bowel inflammation in mice fed ad libitum. These findings suggest that intermittent fasting can reduce inflammation by changing the gut bacterial populations in a beneficial manner. Perhaps in the future, patients with inflammatory gut disorders will be treated by transplantation of feces (or healthy bacteria) from people on intermittent fasting.

A study of overweight asthma patients showed that intermittent fasting can improve their symptoms. The patients were put on a rigorous intermittent fasting regimen in which every other day they consumed only 350 calories but ate as they liked on the intervening days. They were on the intermittent fasting regimen for two months. Before they started the intermittent fasting period and every two weeks during that period, a doctor evaluated the patients' symptoms and respiratory air flow, and blood samples were collected. The patients' airflow improved within two weeks and continued to improve through two months. The levels of several pro-inflammatory cytokines in the blood decreased dramatically during the intermittent fasting period. This is a striking example of intermittent fasting's ability to reduce inflammation and associated clinical symptoms in humans with a disorder caused by tissue inflammation. Subsequent studies have also shown that calorie restriction can reduce other types of inflammation in humans. For example, in a study of more than 200 healthy adults Luigi Fontana found that daily calorie restriction reduced levels of TNF, leptin, and circulating lymphocytes compared to the non-calorie-restricted control group. Moreover, he showed that the daily calorie restriction did not adversely affect immune system responses to vaccines and did not increase the incidence of infections during a two-year period.

OPTIMUM HEALTH: CYCLING BETWEEN STRESS RESISTANCE AND GROWTH MODES

As research on intermittent fasting advanced during the first two decades of the twenty-first century, it became clear that lifelong health can be optimized by relatively short periods of fasting and exercise followed by recovery periods of eating, resting, and sleeping.

In 2018, two articles were published describing the salient data supporting this hypothesis: "Intermittent Metabolic Switching, Neuroplasticity, and Brain Health" by me and a few of my colleagues and "Flipping the Metabolic Switch: Understanding and Applying the Health Benefits of Fasting" by Stephen Anton and his colleagues. The concept draws heavily on the evolutionary considerations described in chapter 1. When food is scarce, cells and organisms conserve resources, bolster their resistance to stress, and do not grow. Upon food acquisition, they utilize the energy, proteins, and fats to grow and become stronger. The impact of physical exertion on cells and organ systems can be viewed from a similar perspective. During the exertion—walking, running, or swimming—cells are subjected to metabolic and oxidative stress and function as efficiently as possible to endure that stress. Then during the period of rest and sleep (after the catching and consumption of a prey animal, for example), cells grow and become stronger and more resilient. Like other species of mammals, humans evolved in environments where such cycles of bioenergetic challenge (fasting and physical exertion) and recovery (eating, resting, and sleeping) were the norm. Research has provided a window into how cells and organ systems evolved to respond to these cycles of negative and positive energy balance in ways that optimize their function and stress resistance so as to maximize their abilities to survive and reproduce.

Cells do not grow during the fasting period. They instead conserve resources and activate genes that encode proteins that bolster the cells' resistance to stress (figure 2.4). The ketones produced during fasting are more energy efficient than glucose. Cells reduce their uptake of amino acids during fasting and reduce the production of proteins that would enable growth of the cells. This reduction in protein synthesis is a consequence of suppression of the activity of proteins in what is called the "mTOR (mechanistic target of rapamycin) pathway." Simultaneously, autophagy and DNA repair are stimulated. Moreover, fasting stimulates the production of antioxidant enzymes, thereby bolstering the resistance

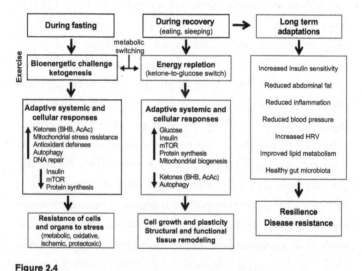

Figure 2.4

Mechanisms by which intermittent fasting improves the function and stress resistance of organ systems throughout the body. BHB = beta-hydroxybutyrate; AcAc = acetoacetate.

to stress. By the time an animal eats after a fasting period, the cells have removed most of their damaged proteins and membranes. They are now "molecularly pristine." The fasting also enhances cells' ability to rapidly take up glucose and amino acids from the blood when food is consumed. The mTOR pathway is activated in response to food consumption, and the cells rapidly synthesize new proteins and membranes, thus enabling them to grow and become physically stronger (as in muscle cells) or functionally more capable (as in brain neurons).

Many of the beneficial responses of cells and organs to intermittent fasting are the same as their responses to exercise. For example, working in my laboratory, the neuroscientist Ruiqian Wan found that every-other-day fasting results in reductions in heart rate and blood pressure as well as in an increase in heart-rate variability. Aerobic-exercise training results in the same changes in the cardiovascular system and by the same mechanism as intermittent fasting—increased activity of the parasympathetic nervous system. Moreover, activation of the parasympathetic nervous system is also responsible for the beneficial effects of regular exercise and intermittent fasting on the gut. Both promote regular bowel movements and so reduce constipation. Similarly, both exercise and intermittent fasting improve glucose regulation, reduce fat accumulation, and improve cognition and mood. In heart, skeletal muscle, liver, and brain cells, regular exercise and intermittent fasting stimulate autophagy, mitochondrial biogenesis, and DNA repair. Beneficial effects of intermittent fasting on the body and brain can be further enhanced by exercise during the fasting period. Both intermittent fasting and exercise slow the aging process and protect against a wide range of age-related diseases.

A question I am often asked after I give a lecture on intermittent fasting and brain health is "What is the best intermittent fasting regimen to counteract aging and promote optimal (brain) health?" I offer two answers. The first answer is that I do not know which regimen is the best because the different regimens have not been compared in the same study. The second answer is that any intermittent fasting eating pattern that results in regular metabolic switching is better than the most common eating pattern of breakfast, lunch, dinner, and an evening snack. In chapters 7 and 10, you will find my advice on how to incorporate intermittent fasting into a lifestyle that also includes a healthy diet (that is, the types of food you consume), exercise, and intellectual challenges.

3 INTERMITTENT FASTING FOR DISEASE PREVENTION AND TREATMENT

Numerous randomized controlled trials of intermittent fasting have been completed in people with obesity. Randomized controlled trials are designed so that equal numbers of people are put in either a treatment group or a control group. In many randomized controlled trials of intermittent fasting, the people in the control group are told not to change their eating pattern and are simply given advice for healthy diets. In other randomized controlled trials, the control group may eat breakfast, lunch, and dinner but will eat fewer calories than normal. An example of a study using a calorie-restricted control group is Michelle Harvie's 5:2 intermittent fasting study in overweight women. Women in the control group consumed 25 percent fewer calories than usual for them at each meal. The configuration of this control group was based on a calculation that their weekly calorie intake would be similar to that of the women in the 5:2 intermittent fasting group, who consumed only 600 calories two days each week. The scientific reason for this study design was to see if intermittent fasting has health benefits beyond what can be accounted for by reduced calorie intake. It does.

There are currently numerous randomized controlled trials of intermittent fasting in progress in patients with various diseases, including diabetes, stroke, multiple sclerosis, epilepsy, kidney disease, and HIV AIDS. Many of these trials can be found at the clinicaltrials.gov website using the search term *intermittent fasting*. A particularly exciting application of intermittent fasting is in the treatment of cancers. At the time of my writing this sentence, trials are in progress in which patients with brain cancer (glioblastoma), breast cancer, lung cancer, prostate cancer, and ovarian cancer are being maintained on an intermittent fasting regimen (5:2 or daily time-restricted eating) during the period when they are being treated with chemotherapy and/or radiation therapy. Other randomized controlled trials of intermittent fasting are being performed in patients at risk for one or more diseases because of the patients' clinical features. For example, intermittent fasting is being tested in patients at risk for cardiovascular disease and stroke because they have high cholesterol levels or high blood pressure. Another example is an ongoing randomized controlled trial designed by me and neurologist Dimitrios Kapogiannis at the National Institute on Aging, in which the participants are at risk for cognitive impairment because of their age (55–70 years old) and metabolic state (obese and insulin resistant). The participants are randomly assigned to the 5:2 intermittent fasting group or the control group, which is provided only with information on healthy diets. Before the participants begin their intermittent fasting and two months after they start, Kapogiannis is evaluating their learning and memory ability and performing functional magnetic resonance imaging (fMRI) to look at activity in neuronal networks throughout their brain. This trial is based on preclinical research in my laboratory demonstrating that intermittent fasting can counteract adverse effects of obesity and insulin resistance on cognition in mice and rats.

The randomized controlled trials of intermittent fasting for specific diseases are based on an understanding of the molecular and cellular abnormalities underlying the diseases, on the one hand, and the impact of intermittent fasting on those abnormalities, on the other hand. This chapter first summarizes what goes wrong in a specific disease and then describes results of animal studies that provide a scientific rationale for randomized controlled trials of intermittent fasting in human patients. Finally, the results of recent trials of intermittent fasting in patients with or at risk for different diseases are presented.

OBESITY AND TYPE 2 DIABETES

The diagnosis of obesity is based on BMI. Someone with a BMI of 30 or more is considered to have obesity. He or she will have high amounts of abdominal fat and so a big waist circumference or "central adiposity." There is compelling evidence that such belly fat promotes tissue inflammation throughout the body. Abdominal obesity also contributes to the insulin resistance and associated poor glucose tolerance in people who are obese. Long-standing obesity greatly increases the risk of diabetes, heart disease, stroke, many types of cancer, kidney disease, and Alzheimer's disease. America is in the midst of an obesity epidemic that affects not only adults but regrettably also many children. Most people my age can attest to the fact that when they were in high school (the early

1970s), very few or no kids were obese in their graduating class. A dramatic increase in obesity has occurred during the past 50 years, and it is strongly linked to increased consumption of simple sugars—particularly high-fructose corn syrup. Also contributing to the epidemic are advances in technologies that have reduced the number of occupations that require substantial physical exertion as well as increased leisure hours spent in front of a screen.

In obesity, fat cells become overloaded with triglycerides, which are used to produce ketones during fasting. But most people with obesity do not fast at all. Recent findings suggest that intermittent fasting not only decreases fat accumulation but can also cause "bad" fat cells to change into "good" fat cells. The abdominal fat in obese people is of a kind called "white fat" that can cause inflammation. In 2017, scientists at the NIH published the results of a study in mice in which they showed that intermittent fasting can cause white fat cells to transform into "beige fat" or "brown fat" cells. It turns out that these brownish cells are beneficial for health because they not only do not cause inflammation but also can "burn" calories to generate heat. Rodents normally have considerable amounts of brown fat in their backs, and those fat cells generate heat to keep their bodies warm when the outside temperature is cold. Humans have small amounts of brown fat in the upper part of their back, but not enough to warm the body. Instead, humans shiver to generate heat. The possibility that intermittent fasting can turn white fat into brown fat suggests that people on intermittent fasting may use those fat cells to generate heat when they are in cold environments.

Numerous randomized controlled trials have shown that adults who are obese lose weight and abdominal fat when they adopt an intermittent fasting eating pattern. These beneficial effects are profound. A recent study in Australia showed that adolescents (12–17 years of age) with obesity were able to comply with an intermittent fasting eating regimen in which they ate only 500–600 calories three days each week while eating regular meals the other four days. Thirty children were enrolled in the study and were on intermittent fasting for 12 weeks. Thereafter, the children were given a choice of continuing intermittent fasting or following a prescriptive eating plan for an additional 14 weeks. Twenty-three of the children chose to continue on intermittent fasting, and all but two of those continued intermittent fasting over the long term. As expected, they lost weight and abdominal fat and exhibited improved mood.

The ability of intermittent fasting to reverse obesity in numerous randomized controlled trials justifies its being prescribed by physicians, together with exercise, as a first-line treatment. Chapter 9 describes how such prescriptions should include patient education, graded transition with flexible stepwise goals, frequent evaluation of progress, and coaching with positive reinforcement.

Unfortunately, the American health care system is currently organized in a manner that discourages physicians from prescribing and following up on specific lifestyle modifications. A patient verging on obesity is often treated in the following way. The patient is seen for 15 minutes at an annual checkup by her primary care physician, who comments that the patient has gained 12 pounds in the past year and now has a BMI of

29. The physician tells the patient that she should try to reduce her weight by cutting back on calories and exercising regularly. The patient is then just scheduled for another checkup in one year, at which time she has gained an additional 10 pounds and has a BMI higher than 30. The doctor has blood work done, and the results indicate that the patient has prediabetes, as indicated by elevated insulin, glucose, and hemoglobin A1c levels. The doctor again simply counsels the patient to lose weight and schedules a follow-up visit six months later. By then, the patient's fasting blood glucose level is 150 milligrams per deciliter (a normal level is 100 milligrams or less), and her high A1c levels confirm that she is diabetic. Based on conversations with a pharmaceutical company representative who periodically visits his office, the physician prescribes a new and expensive diabetes drug. The drug temporarily improves the patient's blood glucose levels but has no impact on her obesity.

The system has failed this patient. The patient was rushed through office visits with no attempt to educate her on how obesity can cause heart disease and stroke, cancers, and even Alzheimer's disease. Had the patient been asked, the physician would have learned that her diet included large amounts of processed foods and soft drinks containing glucose or high-fructose corn syrup. The doctor did not provide the patient with specific prescriptions for changing her eating pattern and diet composition and for incorporating exercise into her daily life. It is this kind of all-too-common scenario that prompted me to devote an entire chapter (chapter 8) to a discussion of a profit-driven health care system that facilitates lifestyles that lead to chronic sickness and early mortality. Although this facilitation may not be intentional, it is true.

People with diabetes have elevated blood glucose levels—hyperglycemia. Diabetes is classified as either type 1 or type 2. In someone with type 1 diabetes, cells in the pancreas fail to produce insulin. Type 2 diabetes results from cells' impaired ability to remove glucose from the blood in response to insulin. Type 1 diabetes is relatively rare and usually affects children whose immune system attacks and destroys the insulin-producing cells in the pancreas. They must then take insulin in highly controlled amounts to keep their blood glucose levels in the normal range. However, they have to be careful not to take too much insulin, or their glucose levels will go too low—hypoglycemia. Type 2 diabetes is increasingly common and usually occurs in adults who are overweight and sedentary. It results from chronic systemic inflammation and oxidative stress, which compromise the ability of muscle, liver, and brain cells to respond to insulin, a situation referred to as "insulin resistance." When insulin binds to its receptor on the surface of a cell, it normally stimulates glucose transporter proteins in the cell membrane. The glucose transporter proteins move glucose from the outside of the cell to the inside, where the cell can then use the glucose to produce ATP. In a person with type 2 diabetes, insulin binds to its receptor, but the receptor is unable to stimulate glucose transport. Because blood glucose levels are abnormally high in this case, the pancreas produces more insulin, and so patients with type 2 diabetes have elevated amounts of insulin in their blood.

According to the US Centers for Disease Control, one in ten Americans is living with type 2 diabetes, and one in three has prediabetes or insulin resistance. Historically, type 2 diabetes affected only adults, but it has now become increasingly common in children. This is very troubling because diabetes is a risk factor for cardiovascular disease, stroke and long-term disability, and early death. Although diabetes most commonly occurs in people who are overweight, it can also occur in people with lower body weights.

It has been established in numerous studies that type 2 diabetes can be prevented and even reversed by caloric restriction and exercise. The results of recent randomized controlled trials suggest that intermittent fasting is also highly effective in restoring cells' ability to respond to insulin, thereby reducing the risk of diabetes in overweight patients and reversing type 2 diabetes in humans. In our studies with Michelle Harvie, 5:2 intermittent fasting increased insulin sensitivity in overweight nondiabetic women. In a randomized controlled trial in Australia, 63 patients with type 2 diabetes who were on 5:2 intermittent fasting for three months exhibited profound reductions in hemoglobin A1c levels. These beneficial effects of intermittent fasting are not immediate. It takes at least two to four weeks to see an improvement in blood glucose regulation. If the person continues his intermittent fasting eating pattern and regular exercise regimen, his glucose regulation will continue to improve and will in many instances become normal. Type 2 diabetes can be reversed with intermittent fasting and exercise. Even in patients who will not or cannot exercise regularly, intermittent fasting can improve and normalize their ability to keep glucose levels within normal limits.

Caution! It is not known whether intermittent fasting is beneficial for people with type 1 diabetes. Because fasting lowers blood glucose levels and increases insulin sensitivity, it would be expected that someone with type 1 diabetes who is on an intermittent fasting eating pattern would require less insulin than if she is eating meals throughout the day. It is therefore possible that intermittent fasting may make this person more prone to hypoglycemia, although this outcome remains to be determined. In fact, when working in my laboratory, the neuroscientist Wenzhen Duan showed that compared to mice fed ad libitum, mice on an every-other-day fasting regimen are more resistant to hypoglycemia when given insulin.

CARDIOVASCULAR DISEASE AND STROKE

Cardiovascular disease is the leading cause of death in the United States, where it claims more than 600,000 lives every year. Stroke is the fourth most common cause of death. Although myocardial infarctions and strokes are catastrophic events, they are ultimately caused by the same insidious pathological process that occurs in the arteries over many, many years prior to the event. The process involves the progressive accumulation of cholesterol-rich "plaques" on the inside of the artery, resulting in progressive narrowing of the vessel. This narrowing reduces blood flow to the heart or brain tissue and may thereby reduce the functional capability of the organ. The atherosclerotic plaque is a site of inflammation and oxidative stress. This inflamed area in the blood vessel can attract platelets, which can aggregate and form a clot. Whereas formation of blood clots serves an important function in preventing excessive bleeding after an injury, a clot that forms inside a narrowed artery can completely block blood flow. When this happens in a coronary artery, the heart muscle cells to which the artery normally supplies blood may be damaged or die—an event called a "myocardial infarction." Similarly, formation of a clot in an artery in the brain results in a stroke that damages and kills neurons.

Metabolic syndrome (the combination of obesity, insulin resistance, and hypertension) and aging are the major risk factors for cardiovascular disease and stroke. Because intermittent fasting can prevent and reverse obesity and insulin resistance and reduce blood pressure, it would be expected that an intermittent fasting eating pattern will lower the risk of a myocardial infarction or stroke. In a study at the University of Oslo, more than 50 obese patients were maintained on 5:2 intermittent fasting for six months, had regular consultation with a dietician, and then went through a six-month unsupervised maintenance period. The patients lost an average of 20 pounds during the first six months and did not regain the weight during the subsequent six months. Their waist circumference, blood pressure, and triglyceride levels decreased significantly during the one-year study period, and their HDL ("good") cholesterol level increased. Researchers have also found that caloric restriction is highly effective in improving the blood lipid profiles in people who have had a myocardial infarction, thereby reducing their risk for a second heart attack.

Krista Varady performed a study aimed at understanding whether a modified every-other-day intermittent fasting regimen can reduce risk factors for cardiovascular disease in overweight people. Participants in the intermittent fasting group ate only 25 percent of their usual daily calorie intake every other day. On the nonrestricted days, the participants were asked to eat 25 percent more calories than usual. Their total calorie intake during the one-year study period was reduced by 25–30 percent compared to subjects in a control group who did not change their eating pattern or calorie intake. Also included was a group that reduced their daily calorie intake by 25 percent but did not intermittently fast. Participants in both the every-other-day fasting group and the daily calorie-restriction group lost about the same amount of weight and had improvements in cardiovascular risk factors.

Intermittent fasting can also help the heart and brain withstand a myocardial infarction or stroke. Working in my laboratory in the 1990s, postdoc Zaifang Yu determined whether intermittent fasting affects the amount of brain damage and dysfunction in rats subjected to a simulated stroke. The stroke-inducing procedure involves first anesthetizing the rat and then making a small opening in the left carotid artery in the neck, through which a fine thread is gently pushed up the artery until the thread reaches the point where the middle cerebral artery branches off the carotid artery. The thread blocks the flow of blood into the middle cerebral artery, which mimics what would happen when a clot forms in the artery. The middle cerebral artery supplies blood to the motor cortex and the striatum, two brain regions that control movement of the limbs. The blood flow is blocked long enough (90 minutes) to cause death of a large number of brain cells in the motor cortex and striatum on the left side of the brain, resulting in permanent impairment in the rat's ability to move its right front and rear legs. The severity of the motor impairment is determined by measuring the rat's ability to use its front paws and to move around in its cage.

The rats in Yu's study were divided into two groups, one fed every day ad libitum and the other made to fast every other day. After three months on the diets, all rats were subjected to a stroke. One day later their motor function was evaluated, and their brains were examined. The rats in the intermittent fasting group had significantly less post-stroke impairment of their front- and hind-limb function and much less brain damage compared to the rats in the control group. Subsequent stroke studies in mice performed by the neuroscientist Thiruma Arumugam confirmed these findings in rats and set the stage for human studies.

Shortly after I established my laboratory at the National Institute on Aging, I initiated collaborations with Ed Lakatta, a cardiologist. Because we had discovered that intermittent fasting protects the brain against a stroke, I collaborated with Ed to see whether intermittent fasting can also protect the heart against a myocardial infarction. Along with postdoc Ismayil Ahmet, we maintained groups of rats on every-other-day fasting or ad libitum control diets for three months and then induced a simulated myocardial infarction by occluding a coronary artery. Some of the rats were euthanized 24 hours after the myocardial infarction, and others were examined by echocardiography during a 10-week period after the myocardial infarction. One day after the myocardial infarction, the amount of heart tissue damage was 50 percent less in the rats in the intermittent fasting group compared to the damage in the rats in the control group. The echocardiography results showed that rats in the intermittent fasting group had better function (strength) of their left ventricle compared to the function in those in the control group.

Intermittent fasting can lower risk factors for a heart attack and stroke and can also lessen damage to the heart and brain if initiated prior to a heart attack or stroke. But can intermittent fasting improve recovery in someone who has already suffered a heart attack or stroke? Okoshi and colleagues recently reported that intermittent fasting improves recovery from a myocardial infarction in rats. When every-other-day fasting was begun the day after the myocardial infarction, the function of the left ventricle improved over time much more than the function of the same ventricle in rats in the control group. Yuan Hu and colleagues recently published a study in which they maintained rats on intermittent fasting or ad libitum diets for one week after subjecting them to a procedure that reduces blood flow to their entire brain. This is a model of cardiac arrest and causes neurons in the hippocampus to degenerate so that the rats have impaired learning and memory. The Chinese scientists found that learning and memory were significantly better after cardiac arrest in the intermittent fasting group compared to the control group.

As recently as the 1980s, the standard advice given by physicians to patients who suffered a heart attack or stroke was to avoid physical exertion and keep well nourished. We now know that being sedentary after a heart attack or stroke is not good. Exercise enhances recovery and protects against a subsequent heart attack or stroke. Our studies and those of the Chinese and Brazilian scientists suggest that intermittent fasting may also improve recovery in such patients. Randomized controlled trials will be required to determine if intermittent fasting can improve recovery of patients who have suffered a stroke or myocardial infarction.

CANCERS

Major progress has been made in the early diagnosis and treatment of more than a dozen different types of cancer. However, cancer remains a leading cause of death throughout the world, and the incidence of many cancers is increasing as a result of the obesity epidemic. In the United States, nearly 2 million people are diagnosed with a cancer every year, and more than 600,000 of them will die from their cancer. In 1971, Morris Ross and Gerrit Bras published an article in which they reported that the incidence of several different types of cancer were greatly reduced in rats maintained on daily time-restricted feeding during their life. Subsequent studies have shown that intermittent fasting inhibits the growth of tumors in animal models of breast, prostate, colon, liver, pancreatic, and brain cancers. In such models, cancer cells are injected into the mouse—either in the tissue where the cancer normally occurs or under the skin—and the formation and rate of growth of the resulting tumor are measured. The fact that intermittent fasting can prevent cancers from forming de novo and can inhibit the growth of existing cancers in animals is consistent with findings from epidemiological studies in humans. In particular, people who are overweight are at increased risk for many different cancers, and the American Cancer Society has determined that overeating accounts for at least 20 percent of all cancer deaths in the United States.

Damage to DNA caused by free radicals can cause mutations in genes that control cell growth. Such mutations can send the cell down a pathway to becoming cancerous. Animals have evolved a highly efficient mechanism that enables them to selectively destroy cells when their DNA becomes damaged sufficiently to make it likely that the cells will become cancerous. The cells undergo a form of programmed cell death called "apoptosis" in which they shrink and then die. As cells undergo apoptosis, their outer membrane remains intact but changes in a way that signals immune cells to seek them out and gobble them up by a process called "phagocytosis." As you read this sentence, thousands of cells in your body are in the process of apoptosis. This is good. It prevents those cells from becoming cancerous. In a wide variety of cancers, however, several genes that normally regulate apoptosis are either mutated or their expression is greatly increased. One such gene codes for p53, a protein that normally plays a critical role in triggering apoptosis when levels of DNA damage become dangerously high. Mutations in p53 occur in a remarkably large number of cancers. The mutations make the cancer cells resistant to dying even when they have excessive amounts of DNA damage. This resistance creates a problem for chemotherapeutic drugs that act by causing DNA damage because cancer cells with p53 mutations can be resistant to being killed by the drugs.

Currently the most common treatments for cancers are chemotherapy and radiation. Both treatments act by causing severe damage to the DNA of the cancer cells. But the treatments also damage normal cells and can cause apoptosis of stem cells in the bone marrow, gut, and other organs, resulting in major side effects. The treatments can also suppress the immune system. Chemotherapy can damage cells in the brain, resulting in cognitive impairment, or so-called chemo brain. The challenge then becomes finding a treatment that kills cancer cells while protecting normal cells such as neurons.

There has recently been exciting progress in the development of treatments that stimulate immune cells to seek out and kill cancer cells, and such therapies are being used for an increasing number of types of cancers.

Intermittent fasting may prevent the development of cancers in the first place and, if a cancer does develop, help in the killing of cancer cells by drugs and radiation. By inducing a mild adaptive stress response, intermittent fasting enhances cells' ability to repair damaged DNA. But this is not the only way that intermittent fasting can prevent cells from becoming cancerous. A study by Vermeij and colleagues showed that daily time-restricted feeding tripled the lifespan of mice with a genetic deficiency in DNA repair. Because these mice usually die at a very young age from cancers caused by excessive DNA damage, intermittent fasting may prevent the development of cancers by reducing free-radical damage to the DNA.

Another mechanism by which intermittent fasting may prevent cancers is by enhancing the killing of newly formed cancer cells by the immune system. Recent findings from Valter Longo's laboratory suggest that intermittent fasting can bolster immune cells' ability to seek out and destroy cancer cells. He found that by stimulating a type of immune cells called "T-cells," intermittent fasting can enhance the killing of breast cancer and skin cancer (melanoma) cells by chemotherapy. For his studies, Longo developed an intermittent fasting regimen involving five consecutive days of a very low daily calorie intake (10–25 percent of usual calories) every other month. In studies of laboratory animals, Vermeij, Longo, and colleagues found that this intermittent fasting regimen can prevent some types of cancer and improve function of the immune system.

In humans, five cycles of this intermittent fasting regimen reduced risk factors for cancer.

Perhaps the most important reason that cancer cells despise fasting has to do with the fact that many cancers rely heavily on glucose as their main energy source. In the 1920s, Otto Warburg discovered that cancer cells are much more sensitive to being killed by glucose deprivation than normal cells. It turns out that cancer cells use glucose to produce ATP by the inefficient process of glycolysis, whereas normal cells use the more efficient process of oxidative phosphorylation. Moreover, many types of cancer cells cannot readily use ketones as an energy source. During fasting, glucose levels are low, and ketones are elevated, so the energy supply for cancer cells is reduced during fasting. The cancer cells' ability to survive exposure to chemotherapy and radiation is reduced when glucose levels are low. This is why intermittent fasting may enhance the effectiveness of these treatments. Because normal cells use ketones, their energy levels remain high during fasting, so the side effects of cancer treatments on normal cells may therefore be reduced by intermittent fasting.

Billions upon billions of dollars have been spent on the development of new drugs and gene therapy approaches to treat specific types of cancers. Intermittent fasting may be a powerful and inexpensive way to enhance the killing of cancer cells by existing chemotherapeutic drugs and radiation. However, intermittent fasting is not yet commonly applied to the treatment of cancer patients. Many more randomized controlled trials will be required in patients with different types of cancers. Although intermittent fasting has been shown to be beneficial in reducing the growth and enhancing the killing of many types of cancer cells in animal models, it is possible that some human cancers can survive the ketogenic state of fasting. Nevertheless, oncologists are impressed by the compelling data from studies in animals that show that intermittent fasting can prevent the occurrence of cancers during aging and that intermittent fasting can be effective in suppressing the growth of tumors and enhancing chemotherapy's ability to kill cancer cells. Dozens of randomized controlled trials of intermittent fasting in cancer patients are currently in progress. It is likely that such trials will lead to treatments in which certain fasting regimens are combined with drugs and radiation in a manner that kills cancer cells and minimizes side effects.

The National Cancer Institute recommends exercise, moderation in calorie intake, and diets rich in vegetables, fruits, and fiber as hedges against cancer. Chapter 6 describes the emerging evidence that the health-promoting chemicals in vegetables, fruits, coffee, tea, and cacao are those that induce a mild beneficial stress response in cells. Such phytochemicals may prevent cancers from developing by bolstering the cells' defenses against free radicals and enhancing their ability to repair damaged molecules. Some of the phytochemicals that have been reported to inhibit cancers in animal studies are the same ones that we and others have discovered can protect neurons against stress: sulforaphane, which is present in high amounts in broccoli, cabbage, and brussels sprouts; curcumin in turmeric root; and quercetin in capers, onions, and kale. Similar to intermittent fasting and exercise, these phytochemicals enhance the elimination of potentially cancerous cells from the body and create an inhospitable environment for cells that have become cancerous.

ALZHEIMER'S DISEASE

Chapter 2 described how aging adversely affects neurons in the brain and how intermittent fasting can counteract these effects of aging. Evidence is accumulating that points to intermittent fasting's potential to reduce the incidence of Alzheimer's disease, an increasingly common and brutal neurodegenerative disorder.

Worldwide, nearly 50 million people have Alzheimer's disease, a number double what it was 30 years ago. About 6 million Americans currently live with Alzheimer's, and this number is projected to triple by 2050. One in every ten people older than 65 is living with Alzheimer's. A diagnosis of Alzheimer's disease is based on a series of cognitive tests that measure short-term memory. Brain imaging will reveal atrophy of the hippocampus and of the frontal, parietal, and temporal lobes. In addition to experiencing impaired short-term memory, Alzheimer's disease patients often have trouble with reasoning, judgment, and language, and their circadian rhythms may become out of sync such that they sleep during the day and are awake at night. A definitive diagnosis of Alzheimer's disease can be made only with a postmortem analysis of brain tissue to establish that sufficient numbers of amyloid plaques and neurofibrillary tangle-bearing neurons are present.

Alzheimer's most commonly strikes people who are older than 65. Most readers will know a relative or acquaintance who has survived a heart attack or a cancer. If this person reaches his eightieth birthday, he has almost a 50 percent chance of being afflicted with Alzheimer's disease. This is a scary thought. Despite the worsening of general health resulting from excessive calorie intake and sedentary lifestyles in industrialized countries, the number of people reaching their seventh, eighth, and ninth decades of life is increasing because advances in the early diagnosis and treatment of cardiovascular disease, some cancers, and diabetes have enabled people who would have previously died from these diseases when they were in their fifties and sixties to live on.

What happens in the brain in Alzheimer's disease and other dementias is complicated and very much a consequence of the usual aging process (figure 3.1). During aging, neurons experience increased damage as a result of oxidative stress, a reduced ability of their mitochondria to produce energy, and an impaired ability to respond to stress in ways that normally keep the neurons healthy. These adversities of aging can render

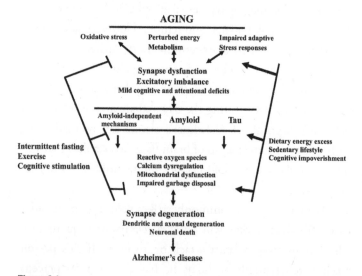

Figure 3.1

The author's vision of how lifestyle factors may either promote or protect against the degeneration of neuronal networks and consequent cognitive impairment in Alzheimer's disease.

neurons vulnerable to excitotoxicity caused by excessive stimulation by the neurotransmitter glutamate. During normal aging, such an excitatory imbalance alone can result in mild deficits in brain function. When combined with the accumulation of the amyloid plaques outside of the neurons and the Tau protein inside the neuron, synapses can degenerate, and the neurons die. Exercise, intermittent fasting, and regular intellectual challenges can protect neurons against the adverse effects of aging on the brain and may prevent Alzheimer's disease.

Soon after my family and I moved to Maryland in 2000, I began noticing that when I talked to my father, who was in Minnesota, on the telephone, he would ask me the same question several times. His short-term memory was failing, and he could not remember that he had already asked me the same question. But he was otherwise doing fine living on his farm and going about his daily routine of taking care of the horses. Four years later he was having difficulty completing his income tax returns and paying bills on time. He was also walking down the quarter-mile driveway to get the mail several times every day, presumably forgetting that he had already retrieved the mail. And yet he was still taking care of himself and regularly driving his car to the grocery store and a friend's house. We made an appointment for him to see Ron Petersen, a neurologist at the Mayo Clinic who is recognized as a leader in research on cognitive impairment and Alzheimer's disease. Petersen was Ronald Reagan's doctor. I had known Ron for many years because we crossed paths at Alzheimer's disease conferences and at NIH scientific review panels. In 2004, when my father was 82, Ron diagnosed him with probable Alzheimer's disease. The word *probable* is used for the initial diagnosis because some people develop cognitive deficits indistinguishable from Alzheimer's disease even though they may have insufficient amounts of amyloid plaques to be ascribed a diagnosis of Alzheimer's disease.

Each year Ron evaluated my father's learning and memory ability and then took an MRI scan of his brain to follow the progression of his brain atrophy. My father's cognitive abilities declined progressively during the next six years, and the size of his hippocampus as well as of the frontal and temporal lobes of his cerebral cortex shriveled. My brother Eric moved back to live with our father during these years, which was invaluable in enabling him to remain in his house on his farm. In the end, we had to place our father in a constant-care facility, and he passed away in his sleep just two months shy of his ninetieth birthday.

At the time of his initial diagnosis, we enrolled our father in the NIH-funded Mayo Clinic Alzheimer's Disease Center, which in addition to clinical evaluations and brain imaging includes a brain autopsy program. Dennis Dickson, the top neuropathologist at the Mayo Clinic, performed the examination of my father's brain tissue and found that amyloid plaque levels were not extensive. However, my father did have massive loss (death) of neurons in his hippocampus, which was almost surely responsible for his having no short-term memory during the last two years of his life. It turns out that about 20 percent of people diagnosed with probable Alzheimer's disease do not have sufficient amyloid plaques to justify a diagnosis of Alzheimer's disease. Many, including my father, are instead diagnosed with what is currently called "hippocampal sclerosis of aging."

There are three known modifiable risk factors for dementia: excessive calorie intake and associated insulin resistance; lack of exercise; and an intellectually unchallenging lifestyle. My father kept physically active until the last five years of his life. Although he never exercised just to exercise, he spent many hours every day walking and doing work around the farm. He was neither overweight nor diabetic. He ate three meals of a reasonably balanced diet every day. However, he did have one risk factor for dementia: he did not keep himself intellectually engaged after he retired at age 60 from his job as an attorney. He did not read books, and his daily activities consisted mainly of routine tasks around the farm. Moreover, when my father was 70 years old, my mother died. Her death of course had a negative impact on his emotional well-being and resulted in a major decrease in his social interactions because he was living alone on the farm.

Intellectual challenges and social interactions reduce the risk for Alzheimer's disease. Research suggests that the reason for this effect can be distilled to the "use it or lose it" concept. Neurotrophic proteins such as BDNF are produced when neurons are active, and their production is critical for the maintenance of existing synapses and the formation of new synapses. Less intellectual and social activity = lack of use of the neuronal networks involved in those activities = less production of neurotrophic proteins = reduction in numbers of synapses = increased death of neurons = poorer brain function. Of course, I did let my father know about our work on intermittent fasting, but by then he was over 80 years old, and there was certainly already considerable loss of synapses and neurons in his hippocampus and elsewhere. The ketone ester described in chapter 5 had not yet been studied. There was nothing that I, Ron Petersen, or anyone else could do to prevent or slow Dad's progressive deterioration into oblivion.

Data from our animal studies, together with associations between obesity, insulin resistance and diabetes, and Alzheimer's disease, and findings from emerging clinical trials suggest a potential benefit of intermittent fasting, in addition to exercise and keeping one's mind intellectually engaged, in reducing the risk for Alzheimer's disease. One study used transgenic mice called "3xTgAD mice" that accumulate amyloid and neurofibrillary tangles in their brains as they age. The mice were maintained on ad libitum, every-other-day fasting, or daily calorie-restriction (40 percent reduction in calories) feeding regimens for one year beginning when they were young adults. A group of normal mice (without amyloid or neurofibrillary tangles in their brains) were also included in the study. At the end of that year, postdoc Veerendra Halagappa tested the learning and memory abilities of the mice. He found that the Alzheimer's mice that were on the ad libitum diet were severely impaired compared to the normal, non-Alzheimer's mice fed ad libitum. In stark contrast, the Alzheimer's mice that had been on every-other-day intermittent fasting or daily calorie restriction exhibited learning and memory ability similar to normal mice. The intermittent fasting completely prevented the cognitive impairment in the Alzheimer's mice. The body weights of the mice on intermittent fasting were greater than the weights of the daily calorie-restriction group, but both groups showed similar excellent performance on maze-learning ability. This indicates that in this mouse model of

Alzheimer's disease intermittent fasting had beneficial effects that are independent of weight loss.

One more puzzle in the Alzheimer's scenario is that there are many very old people who are "sharp as a tack" cognitively but have extensive accumulation of amyloid plaques in their brains. Is there a factor or factors in their lifestyle that has increased the resistance of their neurons to being damaged by the amyloid? We do not yet know the answer to this question. Maybe they have kept physically fit and intellectually engaged, or perhaps they have an intermittent fasting eating pattern. How might intermittent fasting protect nerve cell circuits and preserve cognitive function despite the accumulation of amyloid? The answer may include one or more of the effects of intermittent fasting on the brain shown in figures 2.4 and 3.1. Studies have shown that several of the neurotrophic factors increased in the brain in response to intermittent fasting can protect cultured neurons from being damaged and killed by the Alzheimer's amyloid protein. In addition, whereas the amyloid protein can cause neuronal network hyperexcitability, intermittent fasting can prevent such hyperexcitability.

PARKINSON'S DISEASE

Approximately 500,000 Americans are currently living with Parkinson's disease. Twenty years from now there will be more than 1 million. The symptoms of Parkinson's disease include a tremor in the hands when the arms and hands are at rest, slowed body movements, and muscular rigidity. At autopsy, someone with Parkinson's disease who died will have extensive death of neurons located in a brain region called the "substantia nigra." Those neurons use the neurotransmitter dopamine. Positron emission tomography (PET) brain imaging can measure the amount of dopamine in the brain and thereby provide a measure of how much degeneration of dopamine-producing neurons has occurred in the patient. As the neurons affected in Parkinson's degenerate, they exhibit an abnormal accumulation of the protein alpha-synuclein. In addition to the dopamine neurons, neurons in connected circuits in the cerebral cortex often degenerate, resulting in cognitive impairment. Although it was long thought that dopaminergic neurons are the first to degenerate in Parkinson's disease, we now know that this is not true. Surprisingly, the neurodegenerative process is believed to begin in parasympathetic neurons in the brainstem that innervate the gut. Parkinson's disease patients will often have a history of chronic constipation because the parasympathetic neurons that normally stimulate bowel movements have degenerated.

The boxer Muhammad Ali, my uncle Bud (my father's brother), and the actor Michael J. Fox are three examples of people with Parkinson's disease. A different factor likely resulted in each one's development of Parkinson's. In the case of Ali, Parkinson's disease almost surely resulted from the many blows to his head suffered during his fights. In the case of my uncle Bud, the cause is unclear. He was diagnosed when he was in his early eighties, and so aging certainly played a role. The vast majority of cases of Parkinson's disease have no known genetic cause and affect people when they are older than 65. Because Fox developed symptoms at a very early age, when he was in his thirties, his Parkinson's likely resulted from a genetic factor. About 5 percent of cases of Parkinson's disease are caused by gene mutations, and people with such mutations usually become symptomatic when they are in their thirties, forties, or fifties.

During the past 30 years, major research efforts by geneticists, neuroscientists, and neurologists have elucidated what goes awry in neurons that causes them to degenerate in Parkinson's disease. Two interrelated abnormalities occur in the neurons: their mitochondria become damaged and dysfunctional, and their ability to remove damaged molecules and mitochondria is impaired. As a consequence, the neurons are unable to produce sufficient amounts of ATP to maintain their function, and they accumulate abnormal amounts of the alpha-synuclein protein. Scientists are working to develop interventions that prevent the mitochondrial dysfunction and alpha-synuclein accumulation.

When working in my laboratory, the neuroscientist Wenzhen Duan found that intermittent fasting can protect dopaminergic neurons against a neurotoxin called MPTP in a mouse model of Parkinson's disease. Subsequent experiments by the neuroscientists Navin Maswood and Don Ingram showed that daily calorie restriction can also protect dopaminergic neurons in a monkey model of Parkinson's disease. MPTP selectively kills the dopamine-producing neurons in the substantia nigra. The beneficial effects of intermittent fasting in slowing the disease process in animal models of Parkinson's disease may result from its ability to protect mitochondria against stress and to stimulate autophagy.

The neurologist Bill Langston discovered MPTP while he was working at a medical center in Santa Clara, California, in 1982. Langston had just seen a young patient who had recently developed tremors. Then another five young patients with the same Parkinson's disease symptoms and from the same community appeared at area hospitals. All six of the patients were heroin users, and Langston and his colleagues later discovered that MPTP was a contaminant in the particular batch of heroin they had used. The mechanism whereby MPTP selectively kills dopaminergic neurons is remarkable. Upon entering the brain, MPTP moves into astrocytes, which are a type of glial cell that is as abundant as neurons. Astrocytes contain enzymes called "monoamine oxidases" that act on MPTP to produce the molecule MPP+. Next, the MPP+ is released from the astrocytes and binds specifically to a protein called the "dopamine transporter," which is present only on the membranes of dopaminergic neurons. Just as it does with dopamine, the dopamine transporter moves MPP+ into the dopaminergic neuron, where it accumulates in large amounts. MPP+ impairs the ability of mitochondria to produce ATP, thereby depleting the neuron of the energy it needs to survive.

Mutations in alpha-synuclein can cause Parkinson's disease at an early age, typically when the affected person is in her forties. We used alpha-synuclein mutant transgenic mice that were known to accumulate alpha-synuclein in neurons in the brain and develop motor dysfunction. Postdoc Kathy Griffioen worked with the scientist Ruiqian Wan in a study in which they implanted transmitters to record heart rate in young normal mice and alpha-synuclein mutant mice. The normal mice and alpha-synuclein mutant mice were each divided into three different diet groups: ad libitum feeding of the usual mouse food, alternate-day fasting, and ad libitum feeding of food with high levels of saturated fat. Many months before the alpha-synuclein mutant mice developed motor symptoms, all except for those in the intermittent fasting group exhibited an elevated resting heart rate, which was exacerbated in the high-fat diet group. Further experiments showed that the alpha-synuclein mutant mice had reduced parasympathetic neuron activity, which results in a reduction in heart-rate variability similar to what had been reported in studies of people with Parkinson's disease.

Our attention next turned to the gut. Because alpha-synuclein accumulates abnormally inside of neurons that innervate the gut, and because Parkinson's patients exhibit inflammation in association with the accumulation of alpha-synuclein, we designed a study to test the hypothesis that chronic mild gut inflammation would accelerate the onset of motor dysfunction in alpha-synuclein mutant mice. We predicted that local inflammation in the gut would trigger the propagation of alpha-synuclein pathology from neurons in the gut to neurons in the brain through the vagus nerve, thereby eventually causing dysfunction of the dopaminergic neurons that control body movements. Postdoc Yuki Kishimoto took on the challenge of designing and performing the study. Immunologist Jyoti Sen suggested a method that Kishimoto could use to induce mild inflammation of the intestines based on previous work done by gastrointestinal researchers who study a disorder called colitis. The method involved administering a chemical called dextran sulfate sodium in drinking water.

Every two weeks Kishimoto tested the abilities of mice in the control and gut inflammation groups to control their body movements. As in lumberjacks' log-rolling contests, Kishimoto would place the mice on an accelerating rod and time how long they could stay on the rod. In another test, Kishimoto put ink on the feet of the mice and let them walk on a paper sheet. He would then measure the distance between footsteps on each side of the body. As the alpha-synuclein accumulated in neurons in the brain of the Parkinson's disease mice, their ability to stay on the rotating rod was reduced, and they developed a shorter and often asymmetric stride length. Kishimoto found that abnormalities in both tests occurred significantly earlier in the alpha-synuclein mutant mice in the gut inflammation group than in the mice in the control group. He then performed a series of analyses of the brain, gut, and blood of the mice and upon analyzing the data came to the conclusion that the gut inflammation accelerated the accumulation of alpha-synuclein in neurons innervating the gut, which in turn accelerated alpha-synuclein accumulation sequentially in the vagus nerve, brainstem neurons, and ultimately dopaminergic neurons in the substantia nigra of the brain. By measuring levels of molecules called "cytokines" that are involved in inflammation, Kishimoto further showed that gut inflammation increased brain inflammation in the Parkinson's disease mice. These findings suggest that poor gut health may be a factor in the development of Parkinson's disease.

Clinical trials of intermittent fasting in patients with Parkinson's disease have not yet been performed. Any such trials of intermittent fasting would ideally be done in patients soon after their diagnosis, while they are otherwise fairly healthy. Patients who have progressed to later stages of Parkinson's disease are unlikely to benefit from intermittent fasting because most of their dopamine-producing neurons have already died. My own view is that intermittent fasting is likely to reduce the risk of Parkinson's disease and may be beneficial for people in the early stages of the disease, for three major reasons. First, we know that intermittent fasting can slow down the aging process and that aging is the most prominent risk factor for late-onset Parkinson's disease. Second, there is evidence that suboptimal metabolic health increases one's risk for Parkinson's disease. Third, results from trials of intermittent fasting in animal models of Parkinson's disease suggest that it can protect dopaminergic neurons against degeneration.

INFLAMMATORY DISORDERS AND INFECTIONS

Most chronic diseases involve inflammation of the tissues affected by the disease. In people with rheumatoid arthritis, the joints are primarily affected. Asthma patients suffer from inflammation of the lung airway tissues. Colitis and Crohn's disease are characterized by inflammation in the intestines. This section describes results from studies of animal models and human patients that have shown that intermittent fasting can reduce tissue inflammation and attenuate the disease process in some of the most common inflammatory disorders.

Approximately 1.5 million Americans currently have rheumatoid arthritis, and most of them are women. Rheumatoid arthritis is an autoimmune disorder in which the immune system attacks the joints, where it mistakenly recognizes certain normal molecules as being foreign (nonself). Rheumatoid arthritis results in progressive deterioration of the affected joints, which is accompanied by severe pain. Some drugs can inhibit immune cells from attacking the joints, but there is as yet no cure. A study performed at the University of Oslo by

Jens Kjeldsen-Kragh and colleagues in the early 1990s showed that when rheumatoid arthritis patients fasted for one week and then stayed on a vegetarian, gluten-free diet thereafter, their symptoms were reduced significantly for at least two years.

Valter Longo's research has provided insight into how intermittent fasting affects the immune system in ways that reduce its attack on normal tissues in autoimmune disorders such as rheumatoid arthritis and multiple sclerosis. The immune cells that seek out foreign molecules in the body are called "T lymphocytes." The T refers to the thymus gland, which is an immune organ located in the neck. When the immune system is functioning normally, the T lymphocytes recognize only molecules of viruses, bacteria, and parasites. In cases of rheumatoid arthritis and multiple sclerosis, however, some of the T lymphocytes recognize normal molecules in certain cells of the body as being foreign. When this happens, the lymphocytes proliferate and send signals to other immune cells, which then accumulate in the affected tissue and cause local inflammation and damage to the normal cells in that tissue. It turns out that the lymphocytes involved in the autoimmune attack require glucose to be fully activated. Chapter 2 described how a protein called "mTOR" plays a key role in cell proliferation and how glucose activates mTOR. Glucose levels and mTOR activity are low during fasting, which suppresses T lymphocyte proliferation. The beneficial effects of fasting in people with arthritis may therefore result from suppression of mTOR in T lymphocytes.

In multiple sclerosis, T lymphocytes attack cells called "oligodendrocytes," a type of cell that wraps around the axons of neurons. Axons with oligodendrocytes wrapped around them are called "myelinated" because there is a high amount of the protein myelin in the oligodendrocytes' membranes. By insulating the axon, the oligodendrocytes increase the velocity of the electrical impulse that moves down the axon when the neuron is active. Neurons in both the brain and the spinal cord are affected in multiple sclerosis, resulting in a range of symptoms related to the function of those neurons, including numbness or tingling, difficulty walking, weakness, dizziness, and fatigue. As is the case with many other autoimmune disorders, women are much more prone to multiple sclerosis than men. Together with the symptoms and evaluation of lymphocytes in the patient's blood, an MRI scan will usually reveal abnormalities in "white matter," which are areas with bundles of myelinated axons in the brain, optic nerve, and/ or spinal cord. About 1 million Americans are currently living with multiple sclerosis. Several drugs have been developed that help suppress the immune system and improve the lives of multiple sclerosis patients, but there is as yet no cure.

Recent findings, however, hint at a potential benefit of intermittent fasting for people with multiple sclerosis. The immune system of mice can be made to attack oligodendrocytes by the injection of a myelin protein into the mice. The mice develop symptoms similar to multiple sclerosis, including weakness and reduced walking ability. Studies have shown that daily caloric restriction and a ketogenic diet are beneficial in such mouse models of multiple sclerosis. In one study, every-other-day fasting completely prevented the multiple-sclerosis-like symptoms in a mouse model. Valter Longo found that intermittent fasting with a low-calorie, low-carbohydrate diet has beneficial effects in a mouse multiple sclerosis model. Moreover, his study showed that intermittent fasting reduces the numbers of activated lymphocytes in the blood and, importantly, promoted remyelination of axons. Several studies of the effects of intermittent fasting or ketogenic diets on multiple sclerosis have been published. In one study, a seven-day partial fast (700 calories per day) reduced numbers of abnormally activated lymphocytes and improved self-reported symptoms and quality of life. Kate Fitzgerald and Ellen Mowry at Johns Hopkins University performed a pilot randomized controlled study of 5:2 intermittent fasting in 36 overweight multiple sclerosis patients. The study included three groups: 5:2 intermittent fasting, daily calorie restriction (22 percent reduction), and no change in diet (the control group). At the end of the two-month study period, the patients in the 5:2 intermittent fasting and daily calorie-restriction diets lost weight and had significant improvements in emotional well-being scores. Further studies will be required to determine whether intermittent fasting restores myelination of axons and provides long-term improvements in symptoms of multiple sclerosis.

Nearly 2 million Americans suffer with either ulcerative colitis or Crohn's disease. In colitis, the tissue of the colon and rectum is inflamed. Neither colitis nor Crohn's is thought to be an autoimmune disorder. People with Crohn's disease have inflammation in one or more regions of the digestive tract. The inflamed tissues exhibit accumulations of a different type of white blood cell called "neutrophils" and macrophages. Colitis is more common in people who eat a Western diet high in saturated fat and simple sugars, which suggests a potential benefit of intermittent fasting. Randomized controlled trials in inflammatory bowel disorders will be required to establish whether intermittent fasting is beneficial and which intermittent fasting regimens are the most effective.

As I write, it is June 7, 2021, and more than 600,000 Americans have thus far lost their lives to the coronavirus COVID-19. Patients with COVID-19 most commonly die from respiratory failure as the result of severe inflammation of the lungs. Older age, obesity, diabetes, asthma, and cardiovascular disease are risk factors for morbidity and death in patients with COVID-19. They are also the major risk factors for poor outcomes in patients with influenza. Human studies have shown that intermittent fasting is effective in improving glucose regulation and cardiovascular risk factors as well as in enabling weight loss in overweight people. Intermittent fasting also reduces systemic inflammation and so would be predicted to benefit people who have already contracted COVID-19 or influenza. However, it is not known if and how intermittent fasting might affect the production of antibodies against these viruses that may confer immunity to future exposures to the viruses. Clinical trials of intermittent fasting in people already infected with COVID-19 have not yet been performed, and so at this point medical professionals cannot recommend it as a treatment. Nevertheless, for those who are overweight and insulin resistant, intermittent fasting and exercise can provide protection against a poor outcome should they become infected with the virus.

4 INTERMITTENT FASTING BOLSTERS BRAIN AND BODY PERFORMANCE

Much of the recent flurry of research on intermittent fasting and discussion of intermittent fasting among relatives and acquaintances and on internet forums revolves around its efficacy for losing weight, slowing aging, and treating chronic diseases. This chapter focuses on healthy, normal-weight, physically fit people. Can switching from a typical daily eating schedule of three meals plus snacks to an intermittent fasting schedule improve the performance of their brain and body? If so, what are the changes that occur in cells in response to intermittent fasting that account for the improved performance?

THE IMPORTANCE OF METABOLIC SWITCHING AND KETONES

Fasting is defined by an increase in the concentration of ketones in the blood. The body prioritizes sources of cellular energy, so glucose stored in the liver is used first, lipids (fats) stored in adipocytes (fat cells) are used next, and finally proteins in muscle cells can be broken down to amino acids, which can be used to survive during starvation. In humans who are relatively sedentary, it typically takes about 12 hours to deplete liver glucose stores. Liver glucose will be depleted more quickly in people who exercise during the fasting period. Upon depletion of liver glucose stores, triglycerides in fat cells are broken down into smaller fatty acids, which are released into the blood, from which they are rapidly taken up by liver cells called "hepatocytes." Within the liver cells, the fatty acids are metabolized to three different ketones. A ketone is an organic molecule that has an oxygen atom connected to a carbon atom by a double bond. The three ketones produced during fasting are beta-hydroxybutyrate (BHB), acetoacetate, and acetone. Only BHB and acetoacetate are used as energy sources for cells (figure 4.1). Acetone is not used as an energy source and, in fact, is volatile and expelled in the breath.

As soon as the ketones are produced in the liver, they are released into the blood. The ketones are then actively moved into cells throughout the body and brain by a transport protein in the cell membrane called the "monocarboxylic acid transporter," or MCT. Acetoacetate and BHB move into the cells' mitochondria, where they are used to produce ATP, which is the same energy molecule produced from glucose (figure 4.1). However, two features of BHB and acetoacetate metabolism provide an advantage for cells that glucose does not offer. The first advantage is that, compared to glucose, fewer free radicals are produced during the process of ATP production from the ketones. The second advantage is that ATP production from the ketones is more efficient—more ATP is produced from each ketone molecule than is produced from each glucose molecule.

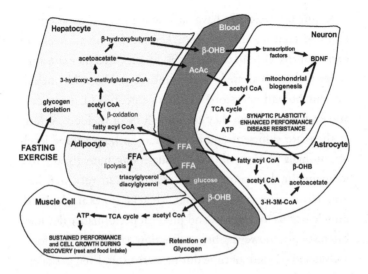

Figure 4.1

During fasting, liver glucose stores (glycogen) are depleted, and blood glucose levels remain low. Next, free fatty acids (FFA) are released into the blood from fat cells. The FFA are then transported into liver cells and converted to the ketones acetoacetate and beta-hydroxybutyrate (BHB). Blood ketone levels rise, and the ketones are transported into cells throughout the body and brain, where they are metabolized to acetyl coenzyme A (CoA), which is then used to produce adenosine triphosphate (ATP). The author's laboratory found that BHB can also stimulate the production of brain-derived neurotrophic factor (BDNF) in neurons. BDNF facilitates learning and memory and protects neurons against stress. Additional benefits of fasting for cells in the brain, heart, and skeletal muscle are the stimulation of mitochondrial biogenesis and autophagy, which effectively remove damaged molecules and increase the number of healthy mitochondria in the cells. TCA = tricarboxylic acid; OHB = hydroxybutyrate.

Humans generally have sufficient lipid stores to enable them to continuously produce ketones for many weeks or even months without eating any food, provided they consume water to keep hydrated. Many animals in the wild (including our human ancestors) experience periods of many days and even weeks when they consume little or no food. As long as there are sufficient amounts of fats to keep ketone levels up, animals in the wild and modern-day humans are able to maintain their muscle mass and function very well in a long-term fasting state. However, when fat stores are depleted, muscle begins to degrade, and physical performance declines. Interestingly, however, brain function is maintained for a considerable length of time during starvation. Even as skeletal muscles, the heart, liver, gut, and other organs shrink during starvation, the brain does not.

Using "ketone strips," some people who have adopted intermittent fasting eating patterns measure ketone levels in a drop of blood obtained from a finger prick. However, the ketone strips are not sensitive enough to detect the increase in ketone levels that occurs early in the fasting period (12–14 hours of fasting). Several companies are therefore working to develop wearable devices to measure low levels of acetone in the breath. At least two companies are selling breath ketone meters. Companies are even working to develop wrist devices that measure acetone released through the skin. Stay tuned . . .

Cells respond to intermittent fasting in ways that enhance their ability to more efficiently acquire glucose and amino acids in the food for energy and protein synthesis, respectively. This makes sense. During the fasting period, the cells prepare themselves such that when food is finally consumed, the cells are able to rapidly ingest and utilize the glucose and amino acids. Proteins in the outer membrane of cells transport either glucose or amino acids into the cell. The glucose transporter is activated by insulin, and that is why insulin reduces the concentration of glucose in the blood—it stimulates the transport of glucose from the blood into cells. This response to insulin is impaired in people with type 2 diabetes, and that is why they are said to have insulin resistance. People who eat three large meals per day plus snacks and do not exercise are highly prone to developing insulin resistance because their metabolic switch is never in the adaptive stress-response mode, so their cells do not experience the metabolic switching required to maintain and enhance their insulin sensitivity and their ability to remove glucose from the blood. As a consequence, these people become insulin resistant and hyperglycemic.

As mentioned previously, during fasting cells reduce their production of new proteins from amino acids and ramp up their removal of damaged molecules. The mTOR protein serves as a switch between the stress-resistance mode of cells during and shortly after exercise and fasting and the growth mode that occurs during the recovery period (figure 2.4). The mTOR protein controls the production of new proteins from amino acids, which is necessary for cell growth. During fasting and exercise, mTOR is turned off, and only new proteins critical for the function and maintenance of the cells are produced; proteins required for growth of the cells are not produced. When mTOR is turned off, a process called "autophagy" (self-eating) is turned on. Autophagy involves the movement of damaged proteins, membranes, and even dysfunctional mitochondria to a specialized organelle (a membrane-bound structure in cells) called the lysosome. In the lysosome are enzymes that break down the proteins into amino acids, which can then be recycled. The overall result of stimulation of autophagy during fasting is to discard and recycle molecular "garbage," thereby making the cell function better.

HORMESIS AND STRESS RESISTANCE

Hormesis refers to adaptive responses of biological systems to moderate environmental or self-imposed challenges through which the system improves its functionality and/or tolerance to more severe challenges.

—Mark Mattson and Edward Calabrese, *Hormesis*

In 2004, I was contacted by Edward Calabrese, a toxicologist at the University of Massachusetts, who wanted to talk with me about how "hormesis" might be applied to understanding and treating brain disorders. Like most scientists at the time, I was unfamiliar with the term *hormesis*. Calabrese's research had shown that many chemicals classified as toxins are actually toxic only at high doses and, more importantly, often have beneficial effects at low doses. Cells can respond to low amounts of certain noxious chemicals in ways that increase their resistance to higher concentrations of the same chemical. The classic example is from murder mysteries, where someone purposely takes low amounts of arsenic or cyanide to build up a tolerance and then gives himself and the person he wants to kill a drink with an amount of arsenic or cyanide that kills the victim but not the murderer. Not being so interested in pursuing research with cyanide and arsenic, I instead decided to test the hypothesis that the health benefits of fasting and exercise involve hormesis.

I found that the beneficial effects of intermittent fasting and exercise *result* from hormesis (figure 4.2). During fasting and exercise, cells in our bodies and brains experience mild stress but become stronger and more resilient for having done so. As described in the preceding two chapters, frequent intermittent periods of fasting can counteract aging and may protect

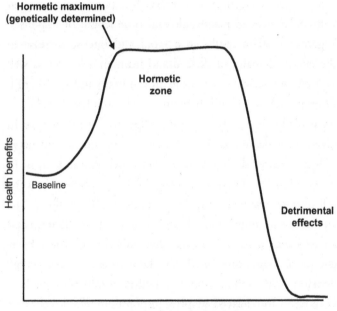

Figure 4.2

Graphic representation of hormesis showing a biphasic curve in which moderate amounts of fasting, exercise, and noxious phytochemicals are beneficial for health, whereas excessive amounts of these environmental challenges can adversely affect health.

against many chronic diseases of the body and brain. Of course, long-term fasting and exercise can be done to excess, resulting in deterioration of muscles and even death. But the good news is that such extremes are never approached with the kinds of very short, frequent periods of fasting that define intermittent fasting and moderate levels of daily exercise.

There are many similarities in how the body and the brain respond to intermittent fasting and exercise. I find it useful to conceptualize intermittent fasting and exercise as different "bioenergetic challenges" to which cells respond similarly, albeit with differences in magnitude and time length of responses. Vigorous exercise results in a rapid and intense increase in the energy demand of skeletal and heart muscle cells as well as increased activity in neurons in the brain and spinal cord. During exercise, blood flow in vessels supplying the latter organs is increased, whereas blood flow to organs involved in digestion is decreased. When activation of muscle and nerve cells is increased during exercise, they experience an increase in sodium ion (Na^+) and calcium ion (Ca^{2+}) movement into the cell through channels in the cell membrane. There also occurs an increase in mitochondrial respiration (ATP production) and an associated increase in oxygen free-radical production. Exercise physiologists discovered that the increased cellular energy demand, Ca^{2+} influx, and free-radical production mediate both rapid and delayed adaptive responses in muscle cells that help them continue to function well during the exercise and then to grow and enhance their resilience during the resting period after the exercise.

The adaptive responses of muscle cells to exercise include the activation of genes that encode proteins that counteract oxidative stress (with antioxidant enzymes), improve mitochondrial health, prevent the accumulation of damaged molecules, and repair damaged DNA. As detailed later in this book, my laboratory found that muscle cells experience adaptive responses to exercise similar to those that occur in brain cells in response to intermittent fasting. Most of our discoveries regarding the mechanisms by which intermittent fasting protects the brain and other organ systems against disease focused on the ability of intermittent fasting to counteract aging and specific disease processes. However, a second approach used to identify brain cells' specific adaptive responses to intermittent fasting drew on the extensive scientific literature on how muscle cells respond to and recover from exercise. Studies had shown that endurance training results in an increase in the number of mitochondria in each muscle cell. In this way, the muscle cells increase their ability to sustain their energy (ATP) levels during extended vigorous exercise. This process is called "mitochondrial biogenesis." Further research identified a protein called "PGC-1α" as a so-called master regulator of the expression of multiple genes required for mitochondrial biogenesis.

In 2007, when neuroscientist Alexis Stranahan was a graduate student in my laboratory, she showed that running-wheel exercise or daily fasting, each by itself, increases the number of synapses on neurons in the hippocampus and that combining running with daily fasting boosts synapse numbers even more than either running or intermittent fasting alone. Moreover, she found that levels of BDNF were increased most with running plus intermittent fasting. Additional experiments by Aiwu Cheng showed that BDNF can cause an increase in the number of mitochondria in neurons. This "mitochondrial biogenesis" is critical for the formation and maintenance of synapses. Because BDNF plays a key role in synapse formation induced by exercise and intermittent fasting, it is likely that exercise and intermittent fasting stimulate mitochondrial biogenesis in neurons.

The synergistic or additive beneficial effects of intermittent fasting and exercise are widespread among organ systems. For example, Krisztina Marosi and Keelin Moehl showed that the ability of daily treadmill training to increase endurance during a two-month period is enhanced in mice maintained on an alternate-day fasting regimen but not in those provided with food continuously. When they also thoroughly analyzed the metabolism of the mice and gene expression in their muscle and liver cells, they discovered that intermittent fasting enhances the production of ketones during endurance training and that intermittent fasting also enhances the ability of the running to stimulate mitochondrial biogenesis in leg muscle cells.

Among endurance athletes, there is now increasing interest in training in a fasted state. Over millions of years of evolution, individuals that were able to sustain endurance in a food-deprived state had a survival advantage. We are now realizing that these evolutionarily potent mechanisms can be used to our advantage should we choose to improve our physical performance. Bodybuilders discovered many years ago that if they forego breakfast, work out in the fasted state midday, and eat during the ensuing 6–8 hours, they lose body fat while still building muscle, thus achieving better muscle definition. Recent controlled studies have shown that intermittent fasting is indeed conducive to the building of muscle in people who do resistance training.

THE BRAIN: SYNAPSES, STEM CELLS, AND NEURONAL NETWORK ACTIVITY

The complex functions of the brain are based on coordinated electrochemical signaling within and between nerve cells (a.k.a. neurons). There are upward of 90 billion neurons in the human brain and more than 100 trillion sites of connectivity or synapses between the neurons. During brain development, neurons arise from stem cells. A neuron begins its life as a spherical cell and then extends several thin processes. One of the processes becomes a long axon, and the others become shorter dendrites. Figure 4.3 is a photograph of a neuron growing on the surface

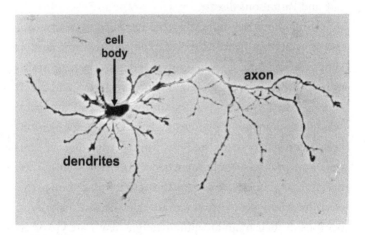

Figure 4.3

A photograph of a single embryonic rat hippocampal neuron that had been growing in culture for three days, during which time a long branching axon and numerous shorter dendrites grew from the cell body. By experimenting with such cultured neurons, the author established fundamental mechanisms by which neurotransmitters and neurotrophic factors regulate the formation of neuronal networks and affect the vulnerability of neurons to degeneration in experimental models of relevance to Alzheimer's disease.

of a culture dish. I took this photograph in 1988. The cell body of the neuron is about five micrometers in diameter, which is one two-hundredth of a millimeter. Three days before I took the photograph, this neuron and thousands of others like it were removed from the brain of a rat embryo and placed in a culture dish. During those three days, the neuron grew its dendrites and axon. In the ensuing days, the axon would continue growing and encounter the dendrites of another neuron with which it could form one or many synapses. Studies of such cultures of neurons have proved valuable in understanding how neuronal networks are formed during brain development as well as what goes wrong in disorders such as Alzheimer's disease and Parkinson's disease.

Throughout the brain, the core neuronal networks consist of excitatory neurons that deploy the neurotransmitter glutamate and inhibitory neurons that deploy the neurotransmitter GABA (figure 4.4). Glutamate is an amino acid that is arguably the most important neurotransmitter in the brain. Without glutamate, the brain is silent. GABAergic neurons keep neuronal network activity from getting out of control. Epileptic seizures occur when GABAergic neurons are not able to properly constrain the activity of glutamatergic neurons. Many people are unaware that the vast majority of neurons in their brain are glutamatergic, fewer are GABAergic, and still fewer use serotonin or dopamine as their neurotransmitter. More people have heard about serotonin and dopamine because of these two neurotransmitters' involvement in depression and addiction, respectively. Serotonin and dopamine "tweak" the activity of glutamatergic neuronal networks, and that is how they cause their effects on mood and behavior.

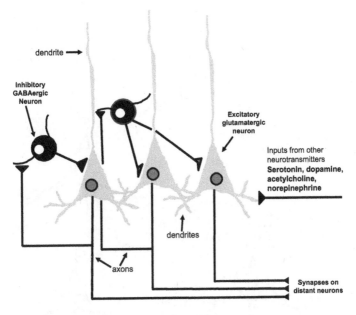

Figure 4.4

Illustration showing the two major types of neurons that form the core neuronal circuitry throughout the brain. Large neurons that deploy the excitatory neurotransmitter glutamate have long axons that project to other neurons within the same brain region or to neurons in other brain regions on the same or opposite side of the brain. Smaller neurons that utilize the inhibitory neurotransmitter GABA have relatively short axons that form synapses on glutamatergic neurons—their activity prevents uncontrolled hyperexcitability of neuronal circuits.

Figure 4.5 is a drawing done in 1988 by Dennis Giddings and conceptualized by my research mentor Stanley Kater when I was working as a postdoctoral scientist in his laboratory in Colorado. The drawing shows the development of a neuron generated from a stem cell and two fates of neurons during aging—either growth and resilience or degeneration. It turns out that many of the same cellular and molecular mechanisms

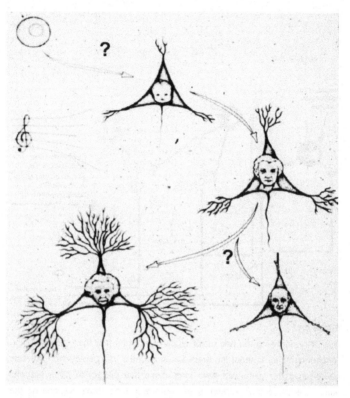

Figure 4.5

Photograph of a drawing conceptualized by Stan Kater in 1988 and done by Dennis Giddings, who was a member of Kater's laboratory when the author was a postdoc there. The drawing shows the development of a neuron generated from a stem cell (*top*) and two fates of the neuron during aging—growth and resilience (*lower left*) and degeneration, as in Alzheimer's disease (*lower right*). The face on the neuron of the healthy adult is that of Kater, and the face on the neuron at the lower left is that of Albert Einstein. The music bar at the left represents environmental inputs such as exercise, intermittent fasting, and intellectual challenges that can determine the fate of neurons during aging.

that control the development of the brain go awry in Alzheimer's disease. Among such mechanisms is the interplay between the neurotransmitter glutamate and the neurotrophic factor BDNF. A proper balance between glutamate and BDNF enables neurons to flourish, whereas too much glutamate and too little BDNF can cause neurons to degenerate.

Much of my research has focused on a brain region called the hippocampus (figure 4.6). The human hippocampus is shaped like a seahorse and exhibits a well-organized neuronal circuitry. The neuroscientist Santiago Ramón y Cajal used a special staining method to visualize neurons and draw their axons and dendrites. In this way, Cajal established the neuronal architecture of many different regions of the brain.

Information from eyes and ears "flows" through nerve cell circuits that converge on the hippocampus, and this is the reason why neuronal circuits in the hippocampus play a critical role in learning and memory. In mice, running-wheel exercise and intermittent fasting increase the number of synapses on dendrites of granule neurons in the hippocampus during a three-month period. Combining intermittent fasting with running results in greater numbers of synapses than do intermittent fasting or running alone. Type 2 diabetes reduces the number of synapses, but intermittent fasting and running-wheel exercise increase the number of synapses on hippocampal neuron dendrites of diabetic mice. Both intermittent fasting and running also stimulate an increase in the amount of BDNF in the hippocampus, suggesting that this neurotrophic factor stimulates the formation of new synapses in the intermittent fasting and running mice. The bottom line is that, at least in mice, intermittent fasting and exercise

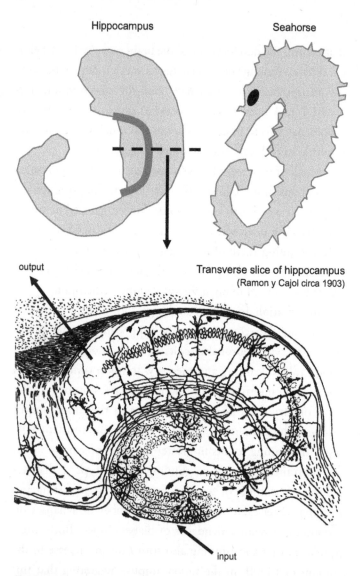

Hippocampus

Seahorse

output

Transverse slice of hippocampus
(Ramon y Cajol circa 1903)

input

Figure 4.6

The human hippocampus is shaped like a seahorse (*top*), and when sliced transversely and examined under the microscope, it exhibits a well-organized neuronal circuitry (*bottom*). The arrows show how information flows into, through, and out of the hippocampus. The drawing of the transverse slice was done by the famous Nobel Prize–winning neuroscientist Santiago Ramón y Cajal in 1903.

can stimulate the formation of new synapses and can reverse adverse effects of diabetes on the brain.

There are stem cells in the hippocampus that divide and can stop dividing and form new neurons in response to exercise. Henriette van Praag discovered that running-wheel exercise increases the number of newly generated neurons in the hippocampus of rats and mice. The stem cells are located in a region of the hippocampus called the "dentate gyrus," and the neurons produced from the stem cells become dentate granule neurons that integrate into the neuronal circuits. Studies have shown that intermittent fasting also increases the number of newly generated neurons that survive and integrate into the dentate gyrus neuronal circuits. Although the mechanisms by which intermittent fasting and exercise stimulate neurogenesis are not fully understood, evidence from several laboratories suggests an important role for BDNF.

While working as a graduate student in my laboratory, Jaewon Lee tackled a long-standing conundrum in the field of aging research. The stress hormone cortisol had been thought to be a "bad guy" when it comes to health, including brain health. However, it turns out that when rats or mice are maintained on a daily caloric-restriction regimen or intermittent fasting, they have elevated levels of corticosterone, which is the rodent equivalent of cortisol in humans. The question then became what explains the fact that intermittent fasting has beneficial effects on the brain and body and increases lifespan, whereas chronic psychosocial stress and obesity, which also elevate adrenal stress hormone levels, have adverse effects on the brain and body? Upon entering neurons, cortisol can bind to either of two receptor proteins, the glucocorticoid receptor or the mineralocorticoid receptor. Previous studies had shown that chronic activation of the glucocorticoid receptor—as occurs in chronic psychosocial stress or diabetes—can cause atrophy of hippocampal neurons and impaired learning and memory, but activation of the mineralocorticoid receptor does not. Lee found that intermittent fasting affects the responses of neurons to cortisol differently than "bad stress" does: intermittent fasting causes a decrease in the levels of the glucocorticoid receptor but not in the levels of the mineralocorticoid receptor, which remained high. This discovery showed that neurons in the brain can respond in either a detrimental or beneficial way to the same stress hormone.

Ketones are one important factor underlying the enhancement of cognition by intermittent fasting. Using cultured embryonic rat brain neurons, postdoc Krisztina Marosi asked whether BHB might act directly on neurons in ways that make them more resilient and resistant to injury and disease. To mimic fasting, she incubated the neurons in a low concentration of glucose and found that when she then added BHB to the culture medium, the neurons increased their production of BDNF. However, this did not occur when she incubated the neurons with a high concentration of glucose. Contemporaneously, Sama Sleiman, working in Moses Chao's laboratory at New York University, infused BHB into rat brains and found that this also increased BDNF levels. Because BDNF can enhance learning and memory and has therapeutic benefits in animal models of depression and neurodegenerative disorders, it is likely that BDNF contributes to the beneficial effects of intermittent fasting and exercise on mood and cognition as well as to resistance to neurological disorders.

A protein called SIRT3, which is located in the mitochondria of neurons, plays a major role in the adaptive responses of neuronal networks to intermittent fasting. Intermittent fasting stimulates an increase in SIRT3 levels in the hippocampus of mice. By recording electrical activity in individual hippocampal neurons from mice that had been maintained on either ad libitum or intermittent fasting feeding regimens, Yong Liu discovered that intermittent fasting enhances the activity of inhibitory neurons that deploy the neurotransmitter GABA. These inhibitory neurons reduce electrical activity in neuronal networks and function to prevent unconstrained activity, such as that which occurs most dramatically during an epileptic seizure. This effect of intermittent fasting is not immediate and instead requires several weeks. A second conclusion is that SIRT3 is required for the enhancement of GABAergic inhibitory tone. In addition, by evaluating the effects of intermittent fasting on the behavior of the mice, Liu found that intermittent fasting reduces anxiety-like behaviors and that SIRT3 is required for the reduction in anxiety. These findings were remarkable because the notion that mitochondria regulate neuronal network excitability in such a manner was not previously recognized.

That it takes several weeks for neuronal networks in the brain to adapt to intermittent fasting has important implications for people who would like to adopt an intermittent fasting eating pattern. Personal communications from people who have tried intermittent fasting suggest that it takes this long for the initial "side effects" of switching to an intermittent fasting eating pattern—including hunger, irritability, and reduced ability to concentrate during the fasting periods—to dissipate. Increased activity of the neurotransmitter GABA may explain why these side effects disappear within a few weeks of starting intermittent fasting.

THE BODY: STRONGER HEARTS, MUSCLES, AND GUTS

The discoveries of what is happening in brain cells that explains how intermittent fasting improves cognition and protects the brain against injury and disease prompted studies of other organ systems. Many organs are controlled by the autonomic nervous system, including the organs of the cardiovascular and digestive systems. The parasympathetic component of the autonomic nervous system deploys the neurotransmitter acetylcholine to reduce heart rate, whereas the sympathetic component deploys norepinephrine to increase heart rate. It was known that intermittent fasting improves glucose regulation by enhancing insulin's ability to stimulate the movement of glucose from the blood into cells. This effect of intermittent fasting was identical to the beneficial effect of exercise on glucose regulation. Might intermittent fasting also affect the cardiovascular system in a manner similar to exercise? To answer this question, Ruiqian Wan measured the heart rate and blood pressure in rats as they were fed ad libitum and then after they were switched to intermittent fasting. The data were crystal clear—resting heart rate and blood pressure were reduced within the first two weeks on the intermittent fasting diet, dropped further during the next two weeks, and then remained stable. Within two weeks of resuming ad libitum feeding, resting heart rate and blood pressure rose back toward where they were before intermittent fasting. These beneficial effects of intermittent fasting are not long-lasting, just as they disappear in an endurance athlete who stops exercising. Wan also performed stress tests on the rats and found that the heart rate and blood pressure of rats on intermittent fasting recovered more rapidly from the stress compared to rats fed ad libitum.

The cell bodies of the parasympathetic neurons are located in the brainstem just above the spinal cord, and the axons of those neurons, which are located in the vagus nerve, form synapses on heart cells (figure 4.7). The reason why endurance athletes have a low resting heart rate is that they have increased activation of the parasympathetic neurons. Their hearts are also more adaptable to stress. Cardiologists can also calculate heart-rate variability, or the variability in the time interval between individual beats. For example, if you have a heart rate of 60 beats per minute, that does not mean that there is exactly 1.0 second between beats; the interval may vary from 0.8 seconds to 1.2 seconds between beats. Healthy, fit people typically have a high heart-rate variability, which means their heart is very good at changing its rate in response to environmental stressors such as exercise. Sedentary folks have a lower heart-rate variability, and patients with heart failure have a very low heart-rate variability. In rats and humans, intermittent fasting increases heart-rate variability by increasing parasympathetic activity.

Because exercise is a potent stimulus for BDNF production in the brain, BDNF is likely to also play a role in causing the low resting heart rate and blood pressure of endurance

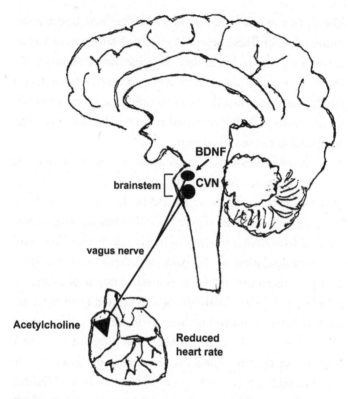

Figure 4.7
Diagram showing neurons in a brain region called the brainstem that send their axons to the heart via the vagus nerve. These cardiovagal neurons (CVNs) are part of the parasympathetic nervous system and deploy the neurotransmitter acetylcholine, which slows heart rate. The author's research has shown that intermittent fasting increases the production of brain-derived neurotrophic factor (BDNF), which increases the activity of the CVNs and thereby reduces resting heart rate and blood pressure.

athletes. By enhancing parasympathetic tone, intermittent fasting and exercise also increase heart-rate variability, which enhances the heart's ability to cope with stress.

In one study, Thiruma Arumugam made an intriguing discovery about how intermittent fasting affects the cardiovascular system in ways that improve its function and stress resistance. He took advantage of a recently refined method for identifying thousands of proteins and for determining whether and to what extent the proteins are phosphorylated on specific amino acids. Mice were divided into four diet groups: a control group fed ad libitum, a group required to fast 12 hours each day, a group required to fast 16 hours each day, and a group on an every-other-day fasting regimen. After three months, the mice were euthanized, and their heart tissue was analyzed. The most prominent effect of all three intermittent fasting regimens was to activate the so-called cyclic guanosine monophosphate (GMP) pathway, resulting in increased phosphorylation of many proteins by cyclic GMP-dependent kinase.

At this point, most of you are probably thinking to yourself, "I have no idea what cyclic GMP is or how it might be related to the beneficial effects of intermittent fasting on heart health." But what if I mentioned a drug named sildenafil? Some of you men may recognize that drug's name, but many others may not. Now what if I told you that sildenafil is marketed under the trade name "Viagra" and that Viagra promotes an erection by increasing cyclic GMP levels in the muscles surrounding blood vessels in the penis so that they become engorged? It turns out that drugs prescribed to increase the blood flow in the coronary arteries of heart disease patients, such as nitroglycerin and sodium nitroprusside, also increase cyclic GMP levels in the smooth muscle cells around the arteries. Although studies aimed at determining whether intermittent fasting might benefit men with erectile dysfunction remain to be performed, it has been shown that intermittent fasting can increase cyclic GMP levels in cells. Moreover, intermittent fasting can improve risk factors for erectile dysfunction, such as obesity and diabetes. Recent clinical trials also suggest sildenafil and drugs that act similarly are effective in treating patients with pulmonary hypertension and heart failure and may also protect heart cells against damage in patients who suffer a myocardial infarction.

Intermittent fasting not only benefits the heart, it can also improve the resilience of skeletal muscles. Without scientific evidence to support the practice, it had become routine for endurance athletes to "load up" on carbohydrates in the days preceding a race and to consume carbs during very long events such as marathons, the Tour de France, and triathlons. But this did not make sense based on the facts that humans evolved to perform physically very well when in a food-deprived state and that ketones, not carbs, enable endurance during periods of fasting. Indeed, working in my laboratory, postdoc Krisztina Marosi and student Keelin Moehl found that mice on intermittent fasting during two months of daily treadmill training were able to run longer and farther than mice that ate whenever they wanted to. There was a strong correlation between blood BHB (ketone) levels and performance—the higher the ketone level, the better the endurance. The fasting mice were using fats and ketones during exercise, whereas the nonfasting mice were using glucose. At the end of the experiment, the mice were euthanized, and their leg muscles were removed and subjected to many different analyses. As expected, muscle cells from mice on intermittent fasting during training were adapted to using ketones. More interesting, however, was evidence that intermittent fasting increased the number of mitochondria in the muscle cells, which likely contributed to improved endurance. Chapter 5 describes how endurance athletes can benefit from competing in a ketogenic state and from taking a ketone ester supplement developed by a colleague of mine at the NIH.

Research suggests that intermittent fasting improves the health of the digestive system. When I was growing up, everyone in my family ate breakfast at about 6:30 a.m., lunch at noon, and dinner at 6:00 p.m. Ice cream or a bowl of cereal at about 8:00 p.m. was also common in our family. With this eating pattern, liver glycogen stores are rarely if ever depleted, and so ketones remain low continuously. I quit eating breakfast in 1982, when I was a graduate student at the University of Iowa, because I was having trouble with chest pain from gastric reflux and found that if I ate breakfast before I rode my bike to the lab, I would have considerable reflux pain. Not eating breakfast reduced the reflux and enabled me to get to the lab early. I have not eaten breakfast since then. In 2013, I was in good running shape and would often run on trails, usually in the late afternoon. It was then that I discovered serendipitously that my running performance was better when I did not eat lunch prior to the afternoon run. Since then, I have skipped breakfast, exercised before eating, and compressed the time window that I consume food to a 6-hour period every day.

Poor intestinal health is common among people in Western societies as a result of their high intakes of calories, saturated fats and sugars, and their sedentary lifestyles. Such people are prone to constipation, inflammatory bowel disorders, and cancers of the stomach and intestines. By increasing parasympathetic tone, intermittent fasting enhances gut motility and promotes regular bowel movements. In addition, emerging evidence suggests that intermittent fasting promotes a healthy gut microbiota. There are more bacterial cells in your intestines than there are cells in the rest of your body, and the types of bacteria in the gut have a major impact not only on the gut itself but also on the rest of the body and even the brain. Nearly 40 years ago, it was found that people with the *Helicobacter pylori* species of bacteria in their stomach and intestines are more likely to develop ulcers compared to people without these bacteria. However, only during the past two decades have methods for rapid large-scale analyses of the "gut microbiota" been applied to understanding how different types of bacteria in the gut affect health in the absence of disease. The composition of gut microbes in healthy and unhealthy people is being established, and fecal transplants are being used to treat certain types of colitis. A fecal transplant involves taking feces from someone with healthy gut microbiota and placing it in the intestines of someone with an illness caused by an unhealthy gut microbiota. However, because a pathogenic (disease-causing) strain of bacteria may be in the feces of the donor, a large effort is being made toward treatment with pure strains or mixtures of strains of healthy bacteria taken in pill form.

Considerable evidence suggests that people who are obese have an unhealthy gut microbiota compared to normal-weight people. Studies have shown that fecal transplants from a normal-weight and healthy donor animal can reverse obesity in rodents. Clinical trials in obese humans are in progress. Guolin Li, Frank Gonzalez, and colleagues at the NIH showed that intermittent fasting reverses obesity in mice, and this effect is associated with a healthier gut microbiome and with a partial conversion of white fat into brown fat. Transplantation of feces from mice on intermittent fasting to obese mice is sufficient to reverse obesity, thus demonstrating a critical role for the gut microbiome. Other studies have showed that a beneficial shift in the gut bacterial species contributes to the reversal of diabetes by intermittent fasting in mice. It remains to be determined whether intermittent fasting without a reduction in overall calorie intake will also improve the gut microbiome, but this seems likely in light of studies showing that intermittent fasting enhances activation of the parasympathetic neurons independently of weight loss. Indeed, reduced parasympathetic nervous system activity has been associated with gut inflammation and an unhealthy microbiome.

Remarkably, the gut bacterial flora can also affect the function of the brain and possibly its susceptibility to disease. Working at the Pennington Biomedical Research Center, Anna Bruce-Keller found that transplantation of gut microbiota from obese mice into normal-weight mice can impair cognition and cause anxiety-like behaviors. Controlled clinical trials will be required to establish if and how the gut microbiome affects brain function in humans. However, one can imagine a future in which "brain-healthy bacteria" are ingested.

MITOCHONDRIA AND CELLULAR RESILIENCE: STRENGTH IN NUMBERS

At an early stage in the evolution of life, a primitive cell was invaded by bacteria, and the cell and the bacteria established a symbiotic (mutually beneficial) relationship in which the bacteria provided energy and the cell in turn provided nutrients for the bacteria. Over millions of years, the symbiotic bacteria evolved into the mitochondria present in the cells of all animals alive today. Mitochondria are sausage-shaped organelles bounded by two membranes. They are present in all of our cells, where their major function is to convert glucose and ketones into ATP, adenosine triphosphate, a molecule that provides the energy required for the survival, growth, and proper function of the cells. In addition to producing ATP, mitochondria have several additional important functions. For instance, they regulate levels of the calcium ion (Ca^{2+}) in the cell, and they often act as an arbiter that determines whether a cell lives or dies by a process called "apoptosis" or "programmed cell death."

Apoptosis normally occurs when a cell is old and worn-out or when it is under severe stress. For example, when a tissue suffers physical damage or ischemia (deficient blood flow), as occurs in a heart attack or stroke, many cells die by apoptosis. When a cell dies by apoptosis, it shrinks and does not "spill its guts" onto its neighbor cells. The apoptotic cell is then recognized and phagocytosed (gobbled up) by immune cells called "macrophages." In this way, apoptosis provides a means of selectively removing a dead cell from a tissue. Events occurring in the mitochondria determine whether a cell undergoes apoptosis when subjected to stress. In animal models of stroke and myocardial infarction, intermittent fasting can protect neurons from dying by apoptosis by bolstering mitochondria.

Intermittent fasting can increase the number of healthy, well-functioning mitochondria in cells. This increase has been shown to occur in skeletal muscle cells and neurons and likely occurs in other types of cells as well. Although much remains to be learned, we now know that major effects of intermittent fasting on mitochondria are accomplished by three general mechanisms. First, intermittent fasting enhances mitophagy, which is the process by which cells remove mitochondria that are not functioning well. Second, intermittent fasting sets in motion the process of mitochondrial biogenesis, whereby healthy mitochondria divide and grow in size. Third, intermittent fasting stimulates the production of the mitochondrial enzyme SIRT3, which improves mitochondrial ATP production, enhances the removal of free radicals, and stabilizes mitochondrial membranes.

Margaret Lewis and Warren Lewis first described the selective removal of damaged and dysfunctional mitochondria by mitophagy more than 100 years ago. If not disposed of in an efficient manner, dysfunctional mitochondria can kill cells or promote cancer because they can generate more than 10 times the amount of free radicals than healthy mitochondria produce. Free radicals can kill cells but can also cause mutations in DNA, and if a mutation occurs in a gene that encodes a protein critical for apoptosis, then the cell may survive and accrue additional mutations that result in uncontrolled proliferation of this type of cell and thus cancer.

Intermittent fasting enhances the mitochondria's ability to maintain their function when the cells in which they reside are subjected to stress. This is best illustrated in tissues that are periodically subjected to bioenergetic stressors such as exercise (muscles and motor neurons) and cognitive challenges (brain). The emerging findings from my laboratory and others suggests that intermittent fasting can bolster the mitochondria's resistance to stress, which may play an important role in protecting against many different diseases.

5 TALES OF A KETONE ESTER

In chapter 2, I pointed out the importance of ketones in some of the benefits of intermittent fasting for the body and brain, including cognition and endurance. You might therefore ask yourself: Can I just ingest the ketones instead of fasting? The answer to this question is: no, not really. It turns out that when the ketones BHB and acetoacetate are ingested in their natural forms, very little of them makes it through the digestive tract and into the blood. The ketones are instead broken down and utilized by microbes in the gut. In fact, gut microbes are thought to consume some of the ketones produced in the liver during fasting. However, ketone levels can be increased somewhat by ingesting so-called medium-chain triglycerides, or "MCTs," and much more by ingesting a chemically modified form of BHB called a "ketone ester." In this chapter, I tell the evolving story of how a biochemist friend of mine at the NIH produced the ketone ester and then collaborated with me and scientists at Oxford University in studies that provided evidence that the ketone ester can enhance endurance and may protect neurons in the brain against Alzheimer's and Parkinson's diseases.

BUD'S IDEA

In the 1970s, Richard "Bud" Veech was working at the NIH to understand the biochemical processes by which ketones are formed in liver cells. He became a leading expert on ketones and cellular energy metabolism. Using his expertise as a biochemist, Bud created a modified form of BHB that could be ingested, absorbed into the blood, and then taken up by cells throughout the body and brain. He developed a chemical reaction that created a molecule called "1, 3 butanediol" that was attached to BHB by a type of bond called an "ester bond." Bud's reason for making this "ketone ester" is that he expected that the butanediol would protect the BHB from being broken down in the gut and so enable the ketone ester to enter the blood. Bud knew that all cells in the body and brain contain enzymes called "esterases" that break ester bonds, so he expected that once the ketone ester was in the body, the esterases would release BHB.

In 2010, Yoshihiro Kashiwaya, a scientist working in Veech's laboratory, found that when rats were fed food containing the ketone ester, the levels of BHB in their blood increased to levels as high as occur during long fasting periods—up to five millimolar. This was the first evidence that BHB levels can be greatly increased without fasting—that is to say, even if liver glucose stores have not been depleted. In contrast, when the rats were fed palm oil, which can be broken down into fatty acid precursors of ketones, BHB levels in the blood were not increased. The reason is that unless glucose is depleted from the liver, the fatty acids are not used to produce ketones and instead are stored in fat cells. Kashiwaya also analyzed brain tissue from the rats and found that the rats fed the ketone ester had more efficient brain cell energy metabolism compared to those not fed the ketone ester.

PROTECTING THE BRAIN AGAINST ALZHEIMER'S DISEASE

Veech knew that I was a leading expert on Alzheimer's disease at the NIH and that I had published a study in 2007 in which we showed that intermittent fasting can preserve learning and memory in a mouse model of Alzheimer's disease. He therefore contacted me to ask whether I would like to collaborate on a study to see whether the ketone ester might be beneficial in our mouse model of Alzheimer's disease. As I mentioned earlier in this book, we had used genetic engineering methods to produce 3xTgAD mice that accumulate amyloid plaques and neurofibrillary tangles in their brains as they age, which resulted in progressive impairment of learning and memory. We decided that Yoshihiro Kashiwaya would come to my laboratory and work with Ruiqian Wan, Eitan Okun, and Mohamed Mughal to perform the experiments. They divided the 3xTgAD mice into two groups: one was fed the ketone ester diet, and the other a control diet that provided the same calories. Mice normally live less than three years. The diets were initiated in middle-aged mice (8–9 months old) at an early stage of the disease when they had some amyloid plaques and neurofibrillary tangles but only slight cognitive impairment. The learning and memory of the mice were tested when they were one year old and again when they were 15 months old. They also tested their anxiety level because my laboratory had previously discovered that the 3xTgAD mice become abnormally anxious.

The learning and memory abilities of the mice were tested in two different ways. One test is called the "water maze test." We have a custom-made swimming pool that is about five feet in diameter and three feet high. The pool contains water that we make cloudy by dissolving a nontoxic white paint in it. A platform of about four inches in diameter is positioned about six inches from the edge of the pool, with the top of the platform submerged about half an inch below the surface. Because the water is murky, the mouse cannot see the platform. To perform the test, a mouse was placed in the pool facing toward the edge opposite the platform. Mice do not like being in the water and so swim randomly to try to find a way out of the pool. The first time a mouse was placed in the pool, it would find the hidden platform by chance. The time it took the mouse to find the platform was recorded. The mouse was then removed from the pool, dried with a towel, and allowed to rest for about ten minutes. The mouse was then placed in the pool again, and the time it took to find the platform was recorded. This process was repeated with each mouse two more times on the same day and four times every day for another five days. Normal mice will use visual landmarks on the walls of the room to remember where the platform is located in the pool, and so the time it takes them to find the platform will become shorter and shorter during the five days of testing. In contrast, as the 3xTgAD Alzheimer's mice become older, their ability to remember the location of the platform will be compromised such that their times to find the platform do not decrease or decrease very little during the five days of testing. On the day after the last test with the hidden platform, the platform was removed, each mouse was placed in the pool, and the amount of time it spent swimming in the area where the platform used to be hidden was determined as another measure of the mouse's memory.

The results were clear. The learning and memory abilities of the Alzheimer's mice were improved by the ketone ester. The Alzheimer's mice that received the ketone ester learned and remembered where the platform was located in the pool significantly more quickly than did the Alzheimer's mice in the control group.

Patients with Alzheimer's disease often exhibit anxiety. The anxiety level of the Alzheimer's mice that did or did not receive ketone ester was evaluated using two different tests, both of them measuring whether the mice were willing to venture into open spaces. If you have ever had mice (or rats) in your house, you likely noticed that they stay close to the walls and do not run across the middle of the room. Wild mice outdoors will also avoid open spaces because they are more likely to be spotted and killed by a hawk, fox, or other predator when they are in an open area. During evolution, individuals that had anxiety that kept them from venturing into open spaces were more likely to survive and pass their "anxiety genes" on to the next generation. Of course, anxiety and fear of potentially life-threatening situations have also been important factors in human evolution. But in modern societies the brain's neuronal networks that control anxiety can become imbalanced in a pathological manner such that they are activated by events (real or imagined) that are not life threatening. In fact, anxiety disorders and the clinical depression that can result from excessive anxiety and worry are the most common mental health disorders in modern societies. Such anxiety disorders are virtually nonexistent in hunter-gatherer societies. In the study of the Alzheimer's mice described here, we found that the ketone ester was very effective in reducing their anxiety.

At the end of the ketone ester study, the mice were euthanized, and their brains were cut into thin sections. As is true with humans with Alzheimer's disease, our 3xTgAD mice exhibited extensive accumulation of amyloid plaques and neurofibrillary tangles in the hippocampus, a brain region that plays a fundamental role in learning and remembering things. To determine whether the ketone ester decreased the amounts of amyloid plaques and neurofibrillary tangles in the hippocampus, we used antibodies against the amyloid and Tau proteins to visualize and quantify the plaques and tangles using a microscope. There were significantly fewer amyloid plaques and neurofibrillary tangles in the brains of the Alzheimer's mice fed the ketone ester compared to those fed the control diet.

The neuroscientist Steve Cunnane used a brain-imaging method called "positron emission tomography" (PET) to look at whether brain cells can use ketones instead of glucose when people are on a ketogenic diet. He found that this was indeed true and that brain cells in Alzheimer's patients can also use ketones. This is important because it has been known for decades that the neurons' ability to utilize glucose is impaired in patients with Alzheimer's. We had discovered in the 1990s that the likely cause of this problem was the amyloid protein, which causes damage to a "glucose transporter" protein that moves glucose across the membrane and into neurons. Cunnane's findings suggest that in Alzheimer's patients the "ketone transporter" protein remains functional, thereby enabling the neurons to function when ketone levels are elevated. This may explain, at least in part, why the ketone ester diet was so effective in protecting neurons against dysfunction and degeneration in our Alzheimer's mice.

During the time of the Roman Empire, it was believed that a person having an epileptic seizure was convulsing because he was possessed by demons. They discovered that if the demon-possessed person was shut in a room for several days and deprived of food, then the person's demonic convulsions would disappear. The scientific reason the "demons" left is likely that during the period of food deprivation, the person's ketone levels increased, and the ketones inhibited the epileptic seizures. Indeed, neurologists who treat patients with epilepsy may prescribe ketogenic diets that contain high amounts of fats and little or no carbohydrates. But, of course, during the Roman Empire it was not known why fasting and ketone bodies inhibit seizures. Recent research has shown that the ketones enhance activity of the inhibitory neurotransmitter GABA, thereby preventing seizures.

Uncontrolled activity in neuronal networks occurs not only in people with epilepsy but also in people with Alzheimer's disease. In fact, people with Alzheimer's disease are much more likely to experience epileptic seizures compared to people who do not have Alzheimer's disease. This "hyperexcitability" results in part from the degeneration of inhibitory GABA neurons in the early stages of Alzheimer's disease. Working in my laboratory, Aiwu Cheng took advantage of genetic engineering and cross-breeding to produce mice that develop epileptic seizures as the amyloid protein begins to accumulate in their brains. The mice eventually die from severe seizures. These mice provided an opportunity to determine just how powerful the ketone ester might be in protecting neurons against Alzheimer's disease. When the mice were fed food containing the ketone ester, they did not develop seizures and did not die. The ketone ester was remarkably effective in preserving the GABAergic inhibitory neurons and thereby keeping neuronal network excitability within normal limits. Clinical trials of the ketone ester in patients with mild cognitive impairment or Alzheimer's disease are currently in progress, and the results are expected to be announced within the next three years.

Moreover, interestingly, dramatic hyperexcitability of neuronal network activity occurs in the brain not only in Alzheimer's disease but also to a lesser extent during normal aging. Lilianne Mujica-Parodi at Stony Brook University recently analyzed brain fMRI data from people of different ages and found that neuronal network activity becomes increasingly destabilized beyond the age of about 50. The study also showed that consumption of a ketone ester and fasting can enhance neuronal network stability. Inasmuch as aging is the major risk factor for Alzheimer's disease and fasting increases ketone levels, Mujica-Parodi's findings are consistent with the possibility that intermittent fasting can reduce one's risk for Alzheimer's disease. Because intermittent fasting and a ketone ester can stabilize neuronal network activity, it would be expected that they will be beneficial for both preventing and treating Alzheimer's disease.

THE YELLOW JERSEY

Bud Veech became aware of the excellent research on endurance training being done by Kieran Clarke and Pete Cox at the University of Oxford. Clarke had begun her research career studying the energy metabolism of heart cells and how the heart responds to stress and ischemia, while Cox studied various aspects of exercise physiology. They came up with a plan to test the possibility that the ketone ester might enhance endurance. In a study of rats, they found that when rats were fed the ketone ester diet, they were able to run 30 percent farther on a treadmill compared to rats in the control group that ate food lacking the ketone ester. They also measured ATP levels in heart cells and found that the ketone ester boosted the cells' energy level. In the same study, the ketone ester enhanced learning and memory, which confirmed our previous findings in the Alzheimer's mice and further suggested the possibility that even the brains of healthy individuals can benefit from elevation of ketone levels.

The next step was to see whether the ketone ester would improve the endurance of human athletes. This is where the story becomes even more interesting. The Oxford scientists performed five separate studies on 39 elite cyclists. In all five studies, they found that the ketone ester improved the performance of the cyclists by providing muscle cells a fuel (BHB) for the efficient production of ATP in their mitochondria. The cyclists taking the ketone ester also had lower levels of lactate in their blood, which is consistent with their ability to go faster and longer without fatiguing. Of course, after the study's results were reported, the cyclists in the study were anxious to know whether they could take the ketone ester (in a liquid form) before and during their races.

Prior to the year 2012, no British rider had ever won the Tour de France. Since then, a British rider has won the Tour six out of nine years and so has worn the yellow jersey. I can tell you with a very high level of confidence that those British riders were using the ketone ester and that many other riders are now using it for the purpose of boosting their endurance. It's OK: the World Anti-Doping Agency approved the ketone ester for use by athletes, which it considers a dietary energy source. Indeed, it is. It is the same ketone (BHB) produced in the body during fasting.

Meanwhile, Kieran Clarke developed a partnership with Geoffrey Woo, who had already become very interested in intermittent fasting and ketones. Woo started a company that produces and sells a drink containing the ketone ester. Brianna Stubbs, who was one of the postdocs involved in the study of the cyclists at Oxford, was recruited to help launch the ketone ester effort. Stubbs is a rower who won two gold medals for Great Britain in the World Rowing Championships in 2013 and 2016. Other companies are also marketing a ketone ester, and many athletes are now using it to boost their performance.

In addition to making determinations as to whether drugs are safe enough for use in humans, the US Food and Drug Administration (FDA) also evaluates the safety of food additives. If it determines that the additive is safe, it gives the additive the "generally recognized as safe" (GRAS) approval. To make this determination, the FDA must have data from studies in humans. A safety study was therefore performed in healthy people who consumed either the ketone ester or placebo for five days, during which time any adverse effects were evaluated. Three different doses of the ketone ester were tested. There were no major side effects with any of the three doses and only mild gastrointestinal symptoms with the highest dose. The FDA therefore assigned the ketone ester GRAS status, which is important because it facilitates the use of the ketone ester in clinical trials. A trial in elderly people with cognitive impairment is in progress as I write.

6 DIET COMPOSITION AND BRAIN HEALTH

In general, studies of the impact of diet on the brain have lagged behind the research on diet and obesity, diabetes, cardiovascular disease, and cancer. One of the major reasons for this is that early epidemiological studies focused only on common health conditions documented in medical records and death certificates. Epidemiological studies entail acquiring data from large groups of people and then analyzing the data to determine whether there are significant associations between two or more factors. For example, data from the Framingham Study that began in Boston in the late 1940s provided some of the earliest evidence that people with high cholesterol levels and/or high blood pressure are more likely to develop cardiovascular disease and die from a heart attack compared to people with lower cholesterol and blood pressure levels. Research in animal models of heart disease and studies that identified genetic mutations that cause early onset atherosclerosis revealed the cellular and molecular events by which high cholesterol levels can cause heart disease. Further studies showed that getting regular exercise, limiting foods rich in cholesterol and saturated fats, and fasting intermittently can reduce levels of LDL ("bad") cholesterol while increasing levels of HDL ("good") cholesterol.

But the results of epidemiological studies are problematic and all too often lead to spurious conclusions regarding the impact of diet on health. The reason is simple. Just because consumption of relatively high amounts of dietary factor X is associated with disease Y does not mean that factor X causes or increases the risk for disease Y. Conversely, just because a low consumption of dietary factor Z is associated with disease Y does not mean that factor Z protects against disease Y. Following the evidence that high blood cholesterol levels are bad for the heart, many people, including scientists studying the issue, concluded that foods that contain cholesterol should be avoided. Eggs contain cholesterol, and so epidemiologists aimed to determine whether people who consume eggs regularly are more likely to develop heart disease. Early studies suggested an association. However, a problem with such studies is that eating the eggs may have nothing to do with risk for heart disease. Instead, people who eat eggs may also have another habit that increases their risk. For example, they may exercise less, smoke more, or eat more red meats, which contain saturated fats, compared to those who eat few or no eggs. Indeed, many studies have shown that saturated fats can elevate cholesterol levels, promote atherosclerosis, and increase the risk of a myocardial infarction.

Genetic factors can confound conclusions. For example, in a hypothetical scenario let's say you and I perform a study of a large population of people distributed throughout New York City aimed at identifying dietary risk factors for hypertension.

After analyzing the data, we find that there is strong correlation between the consumption of turmeric and hypertension. We publish the findings, and several major media outlets cover the findings with headlines such as "New Study Finds That Turmeric Spice Causes High Blood Pressure." However, it turns out that there is a different explanation for the results of our epidemiological study: yes, the diet of Indians (from the country India) includes copious amounts of turmeric root, but studies have determined that Indians have a genetic predisposition to hypertension regardless of their diet.

Myths not supported by data can also emerge. It is likely that when you were a child, your parents told you that "breakfast is the most important meal." Historically, the regular consumption of breakfast occurred as the result of the Agricultural Revolution, when crops were planted and cultivated and meat animals were domesticated. This enabled production of large quantities of grains, eggs, and meat, which were then available in the morning. This daily habit contrasted with that of hunter-gatherers, who had to begin searching for food when they woke up in the morning.

You may be surprised to learn that there is no convincing evidence that not eating breakfast is detrimental to health or brain function. The breakfast food industry will point you to a few studies that concluded that kids who skip breakfast perform more poorly in school. However, the design of those studies was flawed. They took a group of children who normally ate breakfast and divided them into two groups. On the one day allotted to the experiment, half of the children did not eat breakfast, and the other half ate breakfast as usual. Yet we now know that it takes up to one month to adapt to skipping breakfast, during which time the initial hunger, irritability, and reduced ability to concentrate in the morning disappear.

Perhaps more than anyone else, David Allison has studied the literature on breakfast and health. He came to the conclusion that eating breakfast neither promotes nor protects against obesity and related diseases. In 2013, he published the article "Belief beyond the Evidence: Using the Proposed Effect of Breakfast on Obesity to Show Two Practices That Distort Scientific Evidence" in the *American Journal of Clinical Nutrition*. He concluded that many studies on breakfast and health are fraught with two major flaws—the research design lacks probative value, and the reporting of the conclusions are biased. The bottom line on breakfast is that it is not inherently good or bad for health. And as part of an intermittent fasting eating pattern, skipping breakfast can be healthy.

This chapter describes what foods to avoid and which to include in your diet from the perspective of brain health. It turns out that there is nothing particularly unique to the brain when it comes to diet composition. What is bad for the heart is bad for the brain, and what is good for the heart is good for the brain. Of course, there are thousands of books written on diet and health, and some on diet and the brain. Many books focus on a specific dietary component, such as sugar, salt, omega-3 fatty acids, and so on. And bookstore shelves have recently become riddled with "keto" and "Paleo" diet books. But the truth is that the scientific evidence shows that there are three simple principles for healthy eating. First, avoid consumption of simple sugars (glucose, fructose, sucrose), heavily salted foods, fried foods, and highly processed foods. Second, consume a variety of vegetables, fruits, nuts, whole grains, beans, and fish as well as moderate amounts of milk products. Third, cook with extra virgin olive oil.

SAY NO TO SUGARS AND PROCESSED FOODS

A compelling body of research suggests that increased consumption of simple sugars, including glucose, sucrose, and high-fructose corn syrup (HFCS), is very bad for general health and for brain health. Studies of laboratory animals have shown that fructose consumption causes obesity and that the extent of obesity is more than occurs with consumption of the same amounts of glucose or sucrose. Apparently, by mechanisms not yet well understood, fructose reduces metabolic rate, thereby promoting fat accumulation and insulin resistance. There is a remarkable correspondence between the increase in HFCS consumption and the increases in obesity and diabetes in the United States during the past 40 years (figure 6.1). In the recently published book *The Case against Sugar*, Gary Taubes describes the evidence that excess consumption of sugars, including HFCS, is making people obese and sick and that the sugar industries have endeavored to cover up the evidence using methods similar to those employed by the tobacco industry. Here I focus on the lesser-known but profound adverse effects of consuming high amounts of sugars and highly processed foods on the brain.

In 2008, Alexis Stranahan performed a study in middle-aged male rats in which she asked if and how a "fast-food diet" affects the hippocampus. She fed rats a diet high in saturated fat, and they drank water containing HFCS. Control

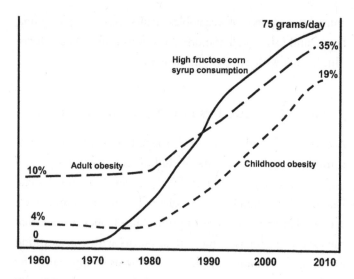

Figure 6.1

The dramatic increases in adult and childhood obesity in the United States are strongly associated with increased fructose consumption during the past 40 years. Based on data from the US Centers for Disease Control.

rats consumed a healthy, mostly vegetarian diet. After eight months on the diets, the rats with excessive sugar intake had gained more weight than did the rats on the control diet. Stranahan then evaluated the learning and memory abilities of both groups of rats using a maze that tests their ability to remember a path they had previously traveled. This spatial navigation requires well-functioning neuronal circuits in the hippocampus. Compared to rats fed a normal balanced diet and plain drinking water, the fast-food diet rats exhibited impaired hippocampus-dependent learning and memory and reduced numbers of synapses on hippocampal neurons. Stranahan also assessed the function of neuronal circuits in the hippocampus by stimulating axons and recording from the neurons onto which the axons were connected by synapses. She found that the ability of hippocampal synapses to "strengthen" in response to vigorous stimulation was impaired in rats on the HFCS and high-fat diet. Moreover, the fast-food diet caused a decrease in the amount of BDNF in the hippocampus, which may explain why learning and memory were impaired.

Scott Kanoski's laboratory at the University of Southern California has shown that learning and memory are impaired in adolescent rats that consume drinking water containing an amount of HFCS (11 percent) that is similar to the HFCS content of widely consumed soft drinks. Moreover, he found that when the young rats consumed the HFCS for only a one-month period, their learning and memory were impaired when tested five months later. If a similar adverse effect of consumption of HFCS on cognition occurs in humans, then this does not bode well for the brain health of children who regularly consume HFCS-containing drinks. Because we are learning more and more about how the diets of pregnant women and their children affect the developing brain, I devote much of the next chapter to that important topic.

Several studies have shown that obesity and diabetes impair cognition in humans, including children. On average, children and adults with obesity perform more poorly in tests of learning and memory compared to those with a normal body weight. For example, Lucy Cheke at Cambridge University tested the memory of 50 young adults (18–35 years of age) with widely varying BMIs, ranging from low normal (BMI of 18) to severe obesity (BMI of 51). She used a computer-based treasure-hunt task that tested the participants' ability to remember what, where, and when from two different scenes in which participants hid food behind objects (one scene on the first day, the other on the second day). They then had to recall which food they hid, where they hid it, and when. Cheke found that there was a significant negative correlation between BMI and memory ability, such that people with higher BMIs performed more poorly than those with low BMIs. Brain-imaging studies have also shown that the size of the hippocampus is smaller in people with abdominal obesity and diabetes than in normal-weight people. Because overconsumption of glucose, sucrose, and HFCS increases the risks for obesity and diabetes, it follows that foods replete with these simple sugars should be avoided in order to keep the brain functioning well.

But sugars are not the only component of processed foods that can be bad for the brain. Others are salt (sodium), saturated fats, and "trans" fats. Too much salt can cause or exacerbate hypertension. Hypertension is a major risk factor for stroke and increases the risk of vascular dementia (memory impairment resulting from "ministrokes") and Alzheimer's disease. There is strong evidence that people who consume high amounts of red meats and fried foods are at increased risk for cardiovascular disease. In the 1980s, this saturated-fat scare resulted in a major effort to use vegetable oils instead of animal fats for cooking. Margarine with vegetable oils was marketed as being healthier than butter. However, the method used to mass-produce vegetable oils and margarine involved "hydrogenation," a process in which hydrogen atoms are added to the carbon chain of the fatty acid, thereby reducing the number of unsaturated bonds in the vegetable oil. These processed fats are called "trans fats." They are most commonly made from corn oil and were widely used without research being done to see whether they were in fact healthier than saturated fats. When studies were done, it turned out they were as bad or worse than saturated fats. Trans fats have fortunately been largely eliminated from processed foods.

But we are still left with the fact that consumption of high amounts of saturated animal fats adversely affects the cardiovascular system and brain. This has been clearly shown in animal studies where rodents are fed diets high in saturated fats such as lard. Consumption of these fats can impair learning and memory and can accelerate the disease process in animal models of Alzheimer's and Parkinson's diseases.

The good news is that we now know that several types of fats are beneficial for general health and brain health. Extra virgin olive oil is one such fat, and fish oil is another. Omega-3 fatty acids are good for both general health and brain health. Omega-3's are present in high amounts in fish as well as in meaningful amounts in certain plants, including Brussels sprouts, flaxseed, hemp seed, and walnuts. Based on our current understanding of fats and brain health, it would seem prudent to use olive oil for cooking, eat the plants that have omega-3's, and substitute fish for red meat.

VARIETY IS THE SPICE OF BRAIN HEALTH

The places in the world where people live the longest and have the lowest occurrence of chronic diseases are those where the diet consists mostly of a wide variety plants, little or no sugars, and little or no red meat: Sardinia, Italy; Okinawa, Japan; Nicoya Peninsula in Costa Rica; Loma Linda, California (specifically Seventh-day Adventists); and Icaria, Greece. Obesity, diabetes, and heart disease are nearly nonexistent in these populations. It should also be noted that people in these so-called Blue Zones do not smoke and are physically active and socially engaged. Although they all consume a mostly vegetarian diet, they vary as to what specific plants they consume. For example, the Sardinians consume a Mediterranean diet. Nearly half of the calories they consume come from whole grains and only about 5 percent from meat (mainly fish and poultry). The Seventh-day Adventists in Loma Linda eat very little meat, and three-quarters of their calories come from beans, vegetables, and fruits. Okinawans consume mainly vegetables, with a type of sweet potato accounting for up to 30 percent of their calories. They also eat fish and small amounts of pork. People in all of the Blue Zones have relatively low-calorie intakes (less than 2,000 calories per day) compared to people in other regions of modern countries (more than 2,500 calories per day).

Studies in which people are fed specific diets and various health indicators are measured support the notion that a mostly vegetarian diet and consumption of fish is best for general health and brain health. A Mediterranean diet has been the most intensively studied diet in such randomized controlled trials. This type of diet consists of plenty of vegetables, fruits, whole grains, beans, and nuts. Olive oil is a primary source of fat, and modest amounts of dairy products, eggs, and meats are consumed. Monica Dinu and her colleagues at the University of Florence recently reviewed findings from epidemiological studies and randomized controlled trials involving a total of nearly 13 million people. They concluded that there is robust significant evidence that a Mediterranean diet reduces the risk of overall mortality, cardiovascular disease, diabetes, neurodegenerative diseases, and some but not all cancers. In addition, interventional studies have shown that diabetes and heart disease can be reversed when patients switch to and maintain Blue Zone–like diets. It remains to be determined whether the beneficial effects of intermittent fasting on general health and the brain can be potentiated by Blue Zone diets, but I would expect this to be true.

The evidence is strong that diets that include a variety of vegetables, fruits, whole grains, nuts, fish, and some milk products (in particular yogurt) are beneficial for brain health. But clear effects of consumption of specific chemical components of such diets on brain function have been established in only a few instances. Moderate amounts of caffeine can enhance cognitive performance. Dietary components that showed promise in animal studies but failed to result in discernable beneficial effects on cognition in humans include omega-3 fatty acids, Gingko biloba, and vitamin D. Because such studies have been relatively short term (e.g., several months), it is possible that when consumed over longer periods, some of these dietary components might slow down the age-related decline in brain function, but this outcome remains to be determined.

There has for decades been much hype about health benefits of "antioxidants." The notion that aging can be slowed and diseases prevented by consuming large amounts of certain chemicals that directly interact with and neutralize free radicals was popularized by Linus Pauling. He was a chemist who was awarded the Nobel Prize for his discovery of the nature of the chemical bond. Pauling suggested, without providing experimental evidence, that the common cold and cancers can be cured by taking high amounts of vitamin C. Subsequent controlled clinical trials proved Pauling wrong. But the antioxidant dietary supplement industry nevertheless took flight, and the health-conscious public was hypnotized by the attractiveness of an easy fix for their health—"Take vitamin C, vitamin E, gingko, acetylcarnitine, and so on, and you will be healthy." Unfortunately, this is not true. People are wasting money on such supplements.

You can also save money by being aware of the fact that an increasing number of dietary supplements are being marketed with the disingenuous pitch that they will improve your memory. A prime example is Prevagen, which is marketed as containing aequorin, a protein present in some species of jellyfish. The truth is, however, that any protein taken by mouth will be broken down into individual amino acids by enzymes in your stomach. This is what will happen to aequorin: it is destroyed in the digestive system. The FDA is beginning to crack down on unfounded health claims regarding such products. The bottom line is that people are wasting their money on these "snake oils." It would be better for their health if they were to use their money to buy healthy foods.

But eating healthy is often difficult for people with low incomes because many brain-healthy foods are more expensive than "brain-wasting" sugars and processed foods. However, it is possible to eat a healthy diet on a tight budget. Indeed, the folks in many of the Blue Zones are far from wealthy. One can follow the lessons from their diets and obtain the bulk of the daily calories in complex carbohydrates from beans, whole grains (oatmeal, whole wheat, etc.), yams, and sweet potatoes. There are also relatively inexpensive fishes available, such as sardines. Onions, garlic, and turmeric can be used to add flavor to vegetable dishes. Olive oil and nuts can be purchased in bulk. Apples, plums, and bananas are excellent but relatively inexpensive fruits. Modest amounts of milk and eggs can be included in such a diet. As to beverages, water and tea or coffee need be the only drinks consumed on most days. People who are fortunate enough to have a yard can grow their own vegetables.

Foraging to collect edible plants and mushrooms is also a good way to get some exercise. It turns out that many of the plants considered weeds are edible. Numerous excellent books and websites are devoted to edible wild plants and mushrooms. I began foraging several years ago and find many edible mushrooms in the state parks and other wooded areas near where I live. Increasing scientific evidence supports the consumption of mushrooms as part of a healthy diet. A surprisingly large number of edible plants grow in our yard and the edges of the adjacent forest, including dandelions, plantains, garlic, onions, chicory, and wild raspberries. But why are such plants beneficial for health?

THE VEGETARIAN'S ADVANTAGE: WHAT DOESN'T KILL YOU

Nutritionists and marketers of dietary supplements have unwittingly claimed that the chemicals in plants (phytochemicals) that are good for health are antioxidants that directly sop up free radicals. Research during the past two decades has debunked this myth by showing that beneficial phytochemicals act by a very different mechanism that is grounded in the fact that chemical defenses are a major mechanism by which plants prevent their overconsumption by animals. In thinking about how to reconcile the clear evidence that diets containing lots of vegetables and fruits reduce the risks for age-related diseases with the failures of clinical trials of antioxidants, I developed a new hypothesis based on the evolutionary relationships between plants and animals.

Two main questions needed to be answered. First, why do plants produce so many different chemicals? Second, how do cells in our body and brain respond to such phytochemicals? It turns out that many of the chemicals in plants, in particular those concentrated in their vital parts—buds, skins, seeds, and roots—serve the function of dissuading insects and herbivorous and omnivorous animals from eating them. Such noxious chemicals are referred to as "natural pesticides" or "insect antifeedants." Prior to the manufacture of man-made pesticides, there was a large natural pesticide or biopesticide industry. The process began with the testing of specific phytochemicals extracted from various plants to determine their ability to repel insects—their "antifeedant potency." But because the plants have food value, it was advantageous for animals to be able to consume the plants without getting sick. Indeed, herbivores and omnivores evolved four major mechanisms to tolerate ingestion of the noxious phytochemicals and thus enable them to consume sufficient amounts of plant parts (fruits, nuts, leaves, roots) to obtain significant amounts of the nutrients the plants contain (carbohydrates, proteins, and fats).

> Most plant defense mechanisms discourage herbivory by deterring feeding and oviposition or by impairing larval growth rather than killing the insect.
>
> —Opender Koul, *Insect Antifeedants*

The first two mechanisms reduce the amount of the phytochemical that reaches the circulation. The noxious phytochemicals have a very bitter taste that if too strong results in avoidance of ingesting that plant material. A good example is that most fruits that have not yet ripened have relatively high concentrations of bitter-tasting chemicals in their skin—think green tomatoes. As the fruit ripens and the seeds inside are ready to be dispersed, the concentration of the noxious chemicals decreases, and the fruit becomes palatable. The second mechanism that protects against overdosing on toxic phytochemicals is vomiting.

The third mechanism involves enzymes in the liver that rapidly degrade and/or modify the phytochemicals so that they are readily eliminated in the urine. These enzymes are called p450 proteins, and they enable animals to eat plants that contain toxic chemicals by preventing the accumulation of the chemicals in the blood and cells of the body and brain. When you eat a plant, these chemicals remain in your system for only a relatively short time period, typically 10 minutes to an hour or so. This short period is important because it prevents the buildup of very high levels of these potentially toxic natural phytochemicals. It also allows time for cells to recover from exposure to the phytochemicals after the vegetables or fruits are consumed. The rapid elimination of these natural biopesticides contrasts with man-made pesticides such as DDT. Because animals were never exposed to DDT over the course of their evolution, they have no liver enzymes capable of removing them. Large amounts of such pesticides therefore accumulate in tissues and can cause illness and death.

Cellular hormesis is the fourth mechanism that mitigates potentially toxic effects of phytochemicals, and it is the mechanism that may explain most if not all of the health benefits of vegetables, fruits, nuts, coffee, tea, and other foods containing phytochemicals. The phytochemicals activate one or more adaptive cellular stress-response signaling pathways. Except for essential vitamins, phytochemicals that have been demonstrated to have health benefits in controlled studies exert their effects by causing a mild stress response in the cells of the animals or humans that consume them. Scientists in my laboratory and others have elucidated some of the ways in which well-known and widely touted and ingested phytochemicals bolster cellular stress resistance.

In 2015, I wrote an article for *Scientific American* entitled "What Doesn't Kill You . . ." The take-home messages are:

> Plants do not have the option of fleeing predators. As a consequence, they have developed an elaborate set of chemical defenses to ward off insects and other creatures that want to make them into a meal.

> Toxins that plants use against predators are consumed by us at low levels in fruits and vegetables. Exposure to these chemicals causes a mild stress reaction that lends resilience to cells in our bodies. Adaptation to these stresses, a process called hormesis, accounts for a number of health benefits, including protection against brain disorders, that we receive [for example] from eating broccoli and blueberries.

> Chemicals that plants make to ward off pests stimulate nerve cells in ways that may protect the brain against diseases such as Alzheimer's and Parkinson's.

Several examples of widely ingested phytochemicals that increase the resistance of cells to disease demonstrate their benefits. Caffeine is perhaps the most widely ingested phytochemical. It's actions on neurons are particularly notable. Caffeine affects several signaling pathways in neurons that enhance alertness/attention and cognition. It activates genes known to be involved in learning and memory and neuronal stress resistance, including those that code for BDNF and PGC-1α, a protein critical for mitochondrial biogenesis. The phytochemicals curcumin and sulforaphane, which are present in high amounts in turmeric root and broccoli, respectively, activate a protein in the cytoplasm of cells called "Nrf2." Nrf2 then moves into the cell nucleus and activates several genes that encode antioxidant enzymes. Resveratrol, a chemical present in relatively high amounts in red grapes, activates the enzyme SIRT1, which in turn regulates genes in ways that protect cells against stress. Onions and many different types of berries contain quercetin, a chemical that can stimulate autophagy. Interestingly, as I have been describing throughout this book, intermittent fasting can also stimulate all of these adaptive stress-response mechanisms.

All of the beneficial phytochemicals have a bitter taste, attesting to their evolutionary role as natural pesticides or antifeedants. Why do you think insects avoid eating coffee beans and tea leaves? The answer is that these plants have high concentrations of caffeine, which is noxious to insects. Caffeine has a very bitter taste, and consuming too much can be toxic or even lethal to humans, as demonstrated by reports of multiple caffeine-overdose deaths. Did you ever try 100 percent cacao? Its bitterness is conferred by phytochemicals such as epicatechins and theobromine that in moderate amounts provide health benefits. However, whereas humans can tolerate consuming considerable amounts of dark chocolate, some animals cannot, most notably dogs, which lack liver enzymes that detoxify theobromine. This is a good example of different animals having different dose-response hormesis curves. In this case, dogs' theobromine curve is shifted to the left compared to that of humans. That is to say, dogs cannot tolerate amounts of this phytochemical that are easily tolerated by humans. Because dogs evolved as carnivores that never consumed plants with theobromine, there was no reason for them to evolve enzymes that detoxify theobromine. I am glad humans evolved as omnivores because I love dark chocolate!

Importantly, plants produce less of the beneficial phytochemicals when they are treated with man-made pesticides. Being attacked by insects causes a plant to produce more of many of the phytochemicals that are beneficial for our health. This lower production of phytochemicals caused by the use of pesticides makes sense: Why would a plant bother launching its own defenses against insects when a man-made chemical does this for them? Therefore, in addition to not containing man-made pesticides, plants grown organically also contain relatively higher amounts of the healthy phytochemicals they produce.

7 LISTEN UP, MOMS AND DADS

The lifestyles of parents can have a dramatic impact on the health of their children. Children whose parents exercise regularly and eat a healthy diet are more likely to adopt such a lifestyle compared to children whose parents are sedentary and consume excessive amounts of high-calorie foods. But recent research even suggests that whether a *prospective* parent has good or poor metabolic health can influence whether that person's future child is less or more prone to obesity, heart disease, and certain brain disorders. One troubling example of this, detailed later in this chapter, is that children born to women who are overweight and insulin resistant during their pregnancy have an increased risk for autism. This knowledge may motivate overweight women to improve their health by regular exercise and intermittent fasting before they conceive and during their pregnancy.

Three quotes from the article "An Evolutionary Perspective on Why Food Overconsumption Impairs Cognition" that I wrote for the journal *Trends in Cognitive Sciences* capture the take-home messages of this chapter.

Of particular concern is the negative impact of metabolic morbidity resulting from overindulgent sedentary lifestyles on the brains of children.

While the impact of sedentary overindulgent lifestyles on brain development and cognitive trajectories in children is clearly an issue of great concern, so too are emerging findings suggesting that the offspring of over nourished mothers and fathers are predisposed to obesity and poorer cognitive outcomes.

Emerging data suggest that overconsumption of high-energy foods by conceiving mothers and fathers increase the risk of poor or suboptimal cognitive outcomes of their offspring.

Hundreds of research studies have established that when developing children experience excessive energy intake—that is, eat way too much and especially eat too much of foods with high amounts of sugar, salt, and fat—and sedentary lifestyles, they are at increased risk for obesity, diabetes, and cardiovascular disease later in life. Emerging findings from studies of laboratory animals and children suggest that consumption of high amounts of sugars and saturated fats are detrimental to the developing brain. During pregnancy, the developing child's brain grows at a remarkably rapid rate, and its overall structure is established by the end of pregnancy (figure 7.1). But the brain continues to grow after birth and through adolescence. A child's metabolic state is critical for optimal brain development and so for academic achievement and acquisition of linguistic and creative abilities. Children who are overweight or obese will have, on average, smaller brains and poorer academic achievement compared to their classmates who are normal weight. Thus, it is important not only that parents provide their children with

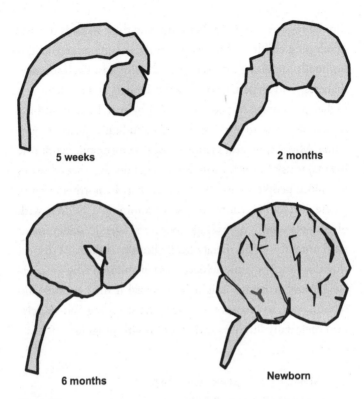

5 weeks

2 months

6 months

Newborn

Figure 7.1
Drawings of the brain of a developing human embryo at the indicated times during pregnancy and at the time of birth. Not drawn to scale.

healthy diets and opportunities for exercise but also that prospective mothers be in a metabolically healthy state during their pregnancy.

The developing brain is highly sensitive to being damaged by lead, mercury, excessive alcohol, and drugs of abuse. Such exposures in utero or after birth disproportionately affect children born to parents of low socioeconomic status, who have limited resources to help them mitigate these hazards to their developing children. But people in these same poorer neighborhoods are also more prone to obesity, insulin resistance, diabetes, and hypertension. In the United States, there is a higher percentage of people affected by metabolic syndrome in several states in the Southeast, including Louisiana, Alabama, Mississippi, and Georgia. Because metabolic syndrome is a major risk factor for stroke and heart disease, these diseases kill more people in southern states than in northern states, where metabolic syndrome is less common. Children with obesity also have, on average, poorer academic achievement than normal-weight, metabolically healthy children. I suspect that the poor academic achievement of children who are overweight and sedentary might be reversed if their parents and schools were to incorporate intermittent fasting and exercise into their daily lifestyles and school health programs.

IMPACT OF METABOLIC HEALTH ON THE CHILD'S BRAIN

Because the childhood obesity epidemic began only about 40 years ago, the impact of being overweight and sedentary on brain development and function has been studied only very recently. Numerous studies have now shown, however, that, on average, children and adolescents who are obese and/or insulin resistant perform more poorly on several tests of cognitive function. For example, Antonio Convit at New York University School of Medicine found that obese adolescents have lower academic achievement and reduced cognitive-processing speed compared to normal-weight peers. MRI analyses of these teens' brains showed that the cognitive deficit is associated with a reduced size of the corpus callosum, where information travels back and forth from one side of the brain to the other. In another study, Convit's team found that working memory (the type of short-term memory concerned with ongoing perceptual and language processing) was poorer in adolescents who are obese. Their poorer working memory was associated with a reduced size of the prefrontal cortex, a region important for decision making. Investigators at Stanford University examined the possible link between insulin resistance and overall brain size in a group of about 50 overweight and depressed youths between 9 and 17 years old. They used MRI scans to measure total brain volume and found that there was a significant association of smaller brain volumes with insulin resistance. This association was independent of age, BMI, socioeconomic status, and depression severity. The Stanford investigators concluded in their article "Insulin Resistance Is Associated with Smaller Brain Volumes in a Preliminary Study of Depressed and Obese Children" that "there may be a significant cost for developing insulin resistance for the developing brain."

Another effect of metabolic syndrome on children and adolescents is an increase in their level of anxiety and their risk for depression. Amy Reichelt at the University of Adelaide recently reviewed published findings from human and animal studies that examined the potential effects of excessive calorie intake and obesity on anxiety and other emotional behaviors. Their major conclusions were that obesity increases the likelihood that children and adolescents will develop a psychiatric disorder, in particular anxiety and depression; among children who are not obese, those with a higher BMI are more likely to develop an anxiety disorder; there is a reciprocal relationship between mood disorders and poor metabolic health such that children who develop an anxiety disorder or depression are more likely to become overweight or obese as adults; and consumption of unhealthy high-calorie Western diets by women before and during pregnancy not only increases their child's risk for obesity and mood disorders but also increases the probability that they (the mothers), too, will experience postpartum depression.

Animal studies have shown that overeating during infancy can adversely affect brain development. For example, Ferreira and colleagues found that when adolescent rats were fed a high-calorie "Western diet" for eight weeks, their learning and memory were impaired, and they exhibited increased anxiety and depression-like behaviors compared to adolescent rats that consumed a healthier amount of calories. Pregnant rats typically give birth to about a dozen or so pups. In one study, the litter size of mother rats was manipulated so that the pups were either overfed (4 pups per nursing dam) or fed less (12 pups per dam). After weaning, all pups were fed a healthy diet, and their learning and memory were tested when they were adults. Two tests of learning and memory were used, one called the "radial-arm maze" and the other called the "novel-object recognition test." In both tests, the rats that were overfed as youngsters performed more poorly than did rats that had received a more normal amount of food as youngsters. Examination of the rats' brains revealed that there was more inflammation in the hippocampus of the rats that had been overfed as youngsters compared to those fed a normal amount. This finding is consistent with evidence from studies of overweight and obese adults showing that excessive calorie intake promotes inflammation in many different organ systems, including the brain.

Beyond inflammation, what might explain the effects of childhood and adolescent insulin resistance and obesity on the structure and function of the neuronal networks involved in cognition and mood? Studies of mice show that overconsumption of saturated fats or sugars or both adversely affects the inhibitory neurons, whose function it is to keep neuronal network activity within normal limits. As described in chapter 4, intermittent fasting and exercise bolster the function and resilience of the inhibitory GABAergic interneurons in a manner that enhances cognition and reduces anxiety. The opposite is true of overeating and obesity, which compromise the function and may even result in the degeneration of the GABAergic neurons. For example, Amy Reichelt at the University of Adelaide fed adolescent rats either a high-fat, high-sugar diet or a control diet and then counted numbers of GABA interneurons in their medial prefrontal cortex, a brain region that plays critical roles in decision making and sociality. Reichelt and her team found that the number of GABAergic neurons was reduced by more than 20 percent in the rats that consumed the high-sugar, high-fat diet compared to those fed the control diet. Moreover, activity of neuronal networks in the medial prefrontal cortex was greatly increased in the young rats fed the unhealthy diet. As described in the next section of this chapter, neuronal network hyperexcitability is a prominent feature of children with autism spectrum disorder (ASD).

The second mechanism by which excessive energy intake—overeating foods high in sugar and fat—adversely affects the structure and function of the developing brain revolves around BDNF. Western diets that are overloaded with sugars and saturated fats reduce the amount of BDNF in the hippocampus and other brain regions involved in learning and memory and the regulation of mood. It is well established that exercise improves mood and enhances cognition, and there is strong evidence from animal studies that these beneficial effects of exercise are mediated by increased BDNF production and consequent changes in both excitatory and inhibitory synapses. Altogether, BDNF affects the structure and function of neuronal networks in a manner that optimizes their functionality and their ability to overcome stress. Intermittent fasting also stimulates BDNF production, and ketones are responsible, at least in part, for the production of BDNF in response to intermittent fasting. As described in chapter 4, BDNF stimulates the formation of new synapses and the production of new neurons from stem cells in the hippocampus. BDNF also bolsters the function of mitochondria and increases the resistance of neuronal networks to stress. In addition, BDNF improves mood and is thought to play a key role in the therapeutic effects of antidepressants (serotonin- and norepinephrine-reuptake inhibitors) and exercise. Moreover, by acting on neurons in the hypothalamus, BDNF can inhibit food intake and thereby protect against overeating and obesity.

Although the recent epidemic of childhood and adolescent obesity is certainly disconcerting, there is good news for parents. The knowledge that a metabolically unhealthy lifestyle that includes the overconsumption of sugars and saturated fats and little or no exercise is detrimental to the developing brains of children and adolescents provides parents with the opportunity to prevent this from happening to their children. Moreover, although prevention is always preferable to amelioration of an existing health problem, we now know that the adverse effects of obesogenic lifestyles on the developing brain can be reversed. Parents should certainly take responsibility for the metabolic health of their children so that they can reach their full intellectual potential and reduce their risk for a mood disorder. Reducing caloric intake, intermittent fasting, and exercise can restore body weight and glucose regulation to normal and can also improve cognition and mood. Eduardo Bustamante, Celestine Williams, and Catherine Davis, the coauthors of a recent article that reviewed the literature on exercise and cognitive and academic performance in youth with obesity, concluded that "regular exercise may be efficacious for improving neurologic, cognitive and achievement outcomes in overweight or obese youth." It will be important to determine whether intermittent fasting can also reverse the adverse effects of being overweight on the bodies and brains of children. It will also be of interest to see whether exercise and intermittent fasting have additive or synergistic beneficial effects as they do in laboratory animals.

PARENTAL METABOLIC HEALTH, EPIGENETICS, AND AUTISM

All the acquisitions or losses wrought by nature on individuals, through the influence of the environment in which their race has long been placed, and hence through the influence of the predominant use or permanent disuse of any organ; all these are preserved by reproduction to the new individuals which arise,

provided that the acquired modifications are common to both
sexes, or at least to the individuals which produce the young.

—Jean-Baptiste Lamarck, translated in *Nineteenth-
Century Science: An Anthology*, edited by A. S. Weber

Jean-Baptiste Lamarck was an eighteenth-century French biol-
ogist and academic whose studies of the relationships between
different organisms led him to a theory of evolution that
included some of the same general conclusions that Charles
Darwin would later establish more clearly and conclusively. But
whereas Darwin concluded that physical and functional adaptations that enable success in survival and reproduction occur
by natural selection over many, many generations, Lamarck
thought that characteristics that parents acquire as a result of
their behaviors and environment can be passed on to their children. For example, if a mother engages in great amounts of
physical labor and thereby develops a large muscle mass, then
her children will also be inclined to have a large muscle mass.

But the notion that characteristics acquired by parents as
a result of their habits and experiences can be passed directly
to their offspring soon faded as the fields of evolution and
genetics took hold. Scientists established that a genetic code
in DNA sequences is translated into amino acid sequences
for proteins. Mutations in this molecular code can result in
either detrimental or beneficial effects on survival and reproduction. Over many generations, the "good" DNA sequences
tend to be retained, and the "bad" sequences tend to be eliminated from the genome. It was clearly established that some
diseases can be inherited as the result of a mutation in a gene
that does not cause early death or infertility. However, during

the past 20 years it has become clear that the inheritance of some physical and behavioral traits cannot be explained by DNA sequences. The field of epigenetics—literally meaning "beyond the genes"—was born. Lamarck's hypothesis has been supported, and some of the underlying molecular mechanisms have been identified.

Epigenetic inheritance can occur as the result of a change in the activation state of a gene and therefore the amount of protein encoded by that gene. Studies of rats and mice have shown that pups born to dams subjected to stress during pregnancy exhibit increased reactivity to stress and a propensity to depression-like behaviors that persist into adulthood. Studies of these animals' DNA have established the effects of maternal stress on the methylation of certain genes—that is, the amount of methyl groups (CH_3) attached to the DNA. The amount of methyl groups on the DNA can determine if and to what extent a gene is turned on. It has been shown that similar effects of maternal stress on DNA methylation occur in humans. Elevated levels of the stress hormone cortisol appear to play an important role in the adverse effects of chronic stress in pregnant females on their offspring. Indeed, studies have shown that elevations of cortisol resulting from chronic psychosocial stress can turn off the BDNF gene.

Chapter 4 considered why some types of transient moderate stress, most notably exercise and intermittent fasting, are good for the brain, whereas chronic psychological stress is bad. Experiencing bad stress during pregnancy may not be good for the developing child's brain, but is being exposed to "good stress" during pregnancy not good, either? For example, let's suppose that Mary and Emma have exactly the same gene for the neurotrophic factor BDNF. Both women are 25 years old and pregnant with their first child. In the preceding 10 years, Mary was sedentary, whereas Emma ran every day. It is known that exercise increases the amount of BDNF in the brain and can thereby improve mood, learning, and memory. Neither woman had family members who suffered from depression. Nevertheless, when Mary's child is a teenager, she develops depression, whereas Emma's child does not. Findings from experiments with laboratory animals and studies of humans suggest that an inherited epigenetic molecular difference in the amount of BDNF protein produced from Mary and Emma's BDNF gene could explain, at least in part, the propensity of Mary's child to depression.

I recall that when I was in high school in the early 1970s, there were no more than two kids with obesity. I also do not recall any kids with autism, nor did I learn about autism from my parents or anywhere else. Even a decade later, the autism prevalence in the United States was reported to be only one in 10,000. Fast-forward thirty years, and the Centers for Disease Control estimated that the prevalence of autism was about one of every 50 children. What explains this dramatic increase? There appear to be several explanations. First, the medical community (psychiatrists) broadened the definition of autism by combining three disorders (pervasive developmental disorders not otherwise specified, Asperger's disorder, and autism disorder) under the umbrella of "autism spectrum disorders." People with an ASD often exhibit social withdrawal and repetitive behaviors that may be accompanied by deficits in language development and cognition. Second, there has been increasing awareness by parents and teachers of the behaviors that suggest an ASD and thus a consequent increase in diagnosis. Third, it has been found that certain environmental factors adversely affect a child's brain development in utero and during the first few years after birth. Here I focus on this third reason for the increase in ASD prevalence.

Considerable emerging evidence suggests that an environmental factor predisposes kids to ASD while they are developing in their mother's womb. It is apparently not true that children with ASD are neurotypical when they are born and then develop an ASD in early childhood. The alterations in the brain's neuronal networks that predispose a child to an ASD are not the result of their being vaccinated or exposed to any exotic chemical in the environment. Instead, maternal obesity and insulin resistance may be major factors that affect brain development in ways that predispose a child to an ASD. In a recent article in *Trends in Neurosciences*, Aileen Rivell and I reviewed the evidence that suggests a major role is played by overconsumption of energy-dense foods and sedentary lifestyles in the dramatic increase in the current prevalence of ASD. The take-home messages are summarized in quotes from the article and illustrated in figure 7.2.

> The rapid increase in the prevalence of autism spectrum disorders (ASD) during the past 40 years is associated with excessive dietary energy intake, particularly fructose, and a concomitant increase in metabolic syndrome (obesity, insulin resistance and hyperlipidemia).

> Children born to mothers with metabolic syndrome and/or diabetes are at increased risk for ASD.

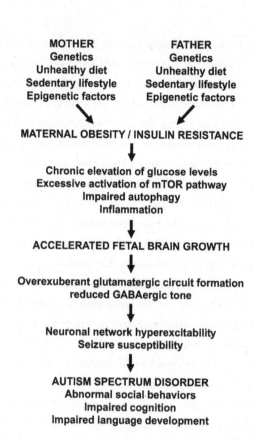

Figure 7.2

How parents' unhealthy diets and lifestyles may increase the risk that their child will have autism.

Studies of humans and animal models suggest that ASD involves accelerated growth of neural progenitor cells and neurons resulting in aberrant development of neuronal circuits characterized by a relative GABAergic insufficiency and consequent neuronal network hyperexcitability.

Genes associated with ASD encode proteins involved in protein synthesis, cell growth and synaptic plasticity, and epigenetic molecular modifications implicated in ASD pathogenesis impact the expression of genes in the same pathways.

Intermittent fasting, exercise, and avoidance of fructose prevent metabolic syndrome, normalize neuronal network excitability and ameliorate ASD-like behaviors in animal models.

Before I describe the salient research findings that led me and Rivell to these conclusions, I want readers to understand that most children born to mothers who are overweight and/or insulin resistant will *not* develop an ASD. This conclusion becomes obvious when one considers that although nearly 50 percent of adults and 20 percent of children in the United States are overweight or obese, only one in 50 children will be diagnosed as being somewhere on "the spectrum." It is also true that many children with an ASD are born to healthy mothers who are not overweight or obese. But being overweight or obese during pregnancy does significantly increase the risk that the child will develop an ASD. Analysis of data from large numbers of people have shown that there is a strong association between the increase in obesity prevalence, the increase in consumption of calories, sugars/fructose, and saturated fats, and the increase in ASD prevalence. Having this knowledge is good news because diet and exercise regimens that prevent and reverse metabolic syndrome in conceiving parents and their offspring may prove beneficial in reducing ASD prevalence and symptom severity.

Although in most instances the cause of each ASD has not been established, ASDs sometimes result from a gene mutation. An unusual feature of many of these mutations is that they occur de novo, which means that the child has the mutation even though neither of his parents has it. This fact suggests that these mutations occurred in the mother's egg, the father's sperm, or very early during development. Many of the mutations are in genes that encode proteins that regulate cellular energy metabolism and protein synthesis, dendrite and axon growth, and learning and memory. Fragile X syndrome is the most common inherited form of ASD and is caused by a mutation in the FMR1 gene on the X chromosome (one of the two sex chromosomes). Mice that are genetically engineered so that they have a dysfunctional FMR1 gene exhibit ASD-like behaviors, including social withdrawal and repetitive movements. Neuroscientists have shown that there is an abnormal increase in neuronal network activity in the hippocampus and other brain regions of the fragile X mice. Hyperexcitability is also evident in neuronal circuits in the brains of a genetic mouse model of Rett syndrome, another ASD. Similar hyperexcitability has been found in studies where MRI was used to evaluate neuronal network activity in the brains of children with an ASD.

A balance in the activities of the excitatory neurotransmitter glutamate and the inhibitor neurotransmitter GABA is required for proper brain functioning. For reasons that are not yet understood, the formation of GABAergic inhibitory synapses is compromised during brain development in ASDs, resulting in hyperexcitability in brain regions that regulate social interactions and cognition. These brain regions include the prefrontal cortex, parietal cortex, and hippocampus. Research suggests that the formation of neuronal networks is accelerated during development of the brain of a child with an ASD. This accelerated growth is even evident by measuring total brain volume in infants with an ASD. Maternal obesity and insulin resistance may accelerate the growth of neural stem cells and of the axons and dendrites of neurons in ASDs. Indeed, animal studies have shown that offspring born to obese and insulin-resistant dams exhibit ASD-like behaviors, including social isolation and repetitive behaviors. It is likely that the accelerated production and growth of neurons during brain development of children of pregnant women who overeat and are sedentary occurs because mTOR is continuously active. These women are not experiencing the periods of fasting and exercise required to turn mTOR off.

Based on the human and animal model studies described in this section of the book, there is reason to believe that parents can reduce chances that their child will develop obesity and/or an ASD by regular exercise, intermittent fasting, and the reduction or elimination of fructose, glucose, sucrose, and saturated animal fats from the diet. It is of course important for general health and mental health that all children get regular exercise and avoid consuming sugars and saturated fats, but several studies have also shown that exercise can alleviate behavioral symptoms in children with ASDs. Studies of overweight pregnant women have shown that exercise is a safe and effective means of improving energy metabolism and preventing abnormal accelerated growth of the child developing in their womb. Experiments in mice in my laboratory at the National Institute on Aging showed that intermittent fasting can reduce anxiety and improve learning and memory by a mechanism involving enhancement of GABAergic tone. Incorporating an intermittent fasting eating pattern into a lifestyle that also includes exercise and healthy foods is something that prospective parents should consider for their own benefit and for the benefit of the family they are planning.

8 BEWARE OF THE "DARK FORCES" AND DON'T EXPECT MIRACLE DRUGS

Intermittent fasting is a lifestyle change that comes at no cost to the "IFer" and has valuable health benefits. These two advantages of intermittent fasting can be viewed as the antithesis to the interests of the pharmaceutical and health care industries, which require people with chronic diseases for their survival. From a not entirely cynical perspective, the ideal scenario for big pharma and the health care industry is large numbers of people with conditions that require long-term treatment with drugs and medical care to reduce symptoms. Certainly, the exponential growth of the pharmaceutical and health care industries in the United States during the past 50 years or so has been bolstered by sedentary, overindulgent lifestyles. In turn, such lifestyles are fostered by the processed-food industry and advances in technologies that eliminate the need for physical exertion. This chapter considers the systemic disincentives for lifestyle modifications that include intermittent fasting eating patterns and examines the potential for the development of drugs aimed at mimicking intermittent fasting in the prevention and treatment of chronic diseases.

The tobacco and processed-food industries and their facilitation of "big agriculture" carry a heavy burden of responsibility for the poor health and premature deaths of many Americans. Moreover, their marketing to foreign countries has and is contributing to increased rates of smoking and consumption of unhealthy foods and drinks in Eastern countries such as Japan and South Korea that previously had very low rates of obesity and high rates of exceptional longevity. From time to time, I have wondered how the leaders and employees of companies that produce and market cigarettes or unhealthy foods are able to rationalize their selling of products that are known to cause diseases such as cancers, cardiovascular disease, and stroke. We are fortunate that in the case of tobacco products the US government established an information campaign and regulations that have reduced the number of Americans that smoke. Meanwhile, though, the epidemics of obesity and diabetes continue unabated.

Although readers are well aware of the advertising practices of companies that produce addictive disease-promoting tobacco and fast-food products, it can be difficult to avoid getting hooked on such products. And in the case of fast foods, customers get a big bang for their buck: a few dollars for a thousand or more calories in one meal. Healthier foods such as fresh vegetables and fruits, nuts, and fish are more expensive than a Big Mac, fries, and a 32-ounce Coke. Two books nicely capture the efforts of large profit-driven companies to suppress or twist the conclusions of scientific research and to dupe the public into buying their products: *The War on Science* by Shawn Otto and *The Attention Merchants* by Tim Wu. The reticence of powerful industries and the politicians they lobby to understand, accept, and rationally apply science to improve the human condition is a big problem. Moreover, major media outlets are also profit driven and have learned that controversy translates into dollars. Coverage of climate change is currently the most prominent example of scientific illiteracy and purposeful avoidance or bending of the truth. Driven by greed and self-preservation, politicians and media outlets sought and found a few rogue "experts" to instill skepticism as to whether human-generated carbon dioxide has been causing global warming and climate change. Similar misinformation campaigns prevent the development and implementation of government regulations on the junk-food industries.

Shown at the center of figure 8.1 are the insidious forces steering many Americans down the pathway to poor health: the processed-food industry, monoculture agriculture, the pharmaceutical industry, and effort-sparing technologies that facilitate sedentary lifestyles. These industries market products that worsen health to the general public, the media, and physicians. But there are ways that people—parents, schools, the government—biomedical research, and health care systems can work together to mitigate or even eliminate the impact of profit-driven, disease-promoting forces. The following measures will be required. Media, parents, and schools should provide clear and frequent information on healthy lifestyles and diets. Governments should restrict advertising of unhealthy foods and should provide incentives for organic farming and disincentives for processed-food production. Biomedical research and physician training have the potential to counteract the negative influences of the big industries on health by emphasizing and implementing lifestyle-medicine

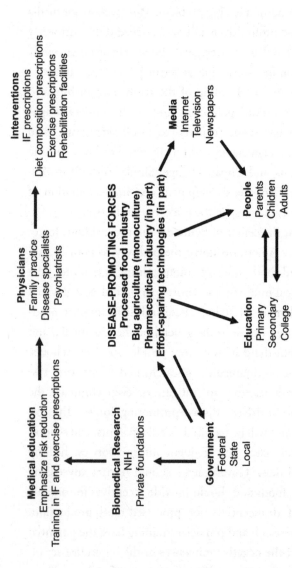

Figure 8.1

Diagram showing how major industries that influence people's lifestyle and medical care decisions promote poor health and how families, education systems, governments, and the biomedical research and physician-training sectors can counteract these industries' negative influences on health (IF = intermittent fasting; NIH = National Institutes of Health).

approaches with the goal of preventing the need for drugs. Prescriptions for intermittent fasting are among such lifestyle-medicine approaches. This will be possible, though, only if health care professionals and social workers are enabled to spend sufficient time and effort with a patient during the time he or she is switching to an intermittent fasting eating pattern.

THE SWEET TOOTH AND FOOD ADDICTION

Nora Volkow is the director of the National Institute on Drug Abuse. During the past two decades, she has performed research studies that show that some of the major changes that occur in the brains of people addicted to drugs also occur in the brains of obese people who have high-sugar, high-salt, and high-fat diets. In the article "The Dopamine Motive System: Implications for Drug and Food Addiction" published in 2017, Volkow reviewed studies that examined the effects on the brain of highly palatable processed foods that are rich in sugar and fats. Her conclusions were as follows:

> In the modern world, conditioning to the many surrounding food stimuli along with the accessibility of reinforcing food can lead to problems. It is presumably the elevation of dopamine by the conditioned stimuli that propels us to go after the chocolate (or other conditioned foods). This effect of conditioned stimuli may help explain the deleterious consequences of enhanced advertising for unnecessary high-calorie foodstuffs (that is, those that are surplus to our daily requirements) as well as the push to manufacture foods that maximize their reinforcing value and hence conditioning effects by mixing ingredients with purposefully calibrated (high) concentrations of fat, sugar and salt. For

foods, enhanced conditioning for a flavour can be associated with the energy content in the food, namely, its value after ingestion. In addition, food manufacturers use all their skills to induce us to eat more of their products (larger portions) than we need, to which we subsequently become conditioned, so that we build up an expectation not only for the high calorie content of the food but also for the delivery of large portions.

Abnormalities in neuronal circuits that deploy the neurotransmitter dopamine play a major role in drug and alcohol addiction. This is also true in people who become dependent on highly palatable obesogenic foods. But in the case of obesity, the brain also has an impaired ability to respond to a hormone called "leptin." Leptin is produced in fat cells and is released into the blood in response to food consumption, much as insulin is released from pancreatic beta-cells when glucose levels rise. Neurons in the hypothalamus at the base of the brain respond to leptin, resulting in the activation of neuronal networks that inhibit food intake and so prevent overeating. In people who are obese, the neurons in the hypothalamus do not respond well to leptin, and so such individuals continue to feel hungry even though they are overeating. This condition is called "leptin resistance" and is analogous to insulin resistance, wherein liver, muscle, and other cells in the body become relatively unresponsive to insulin, and so blood sugar levels remain abnormally high. The good news is that just as exercise and intermittent fasting can prevent and reverse insulin resistance, they can also prevent and reverse leptin resistance.

Another interesting feature of neurons in the hypothalamus is that some of them respond to BDNF, and studies have shown that BDNF suppresses appetite. In fact, mice genetically engineered to produce 50 percent less BDNF than normal overeat and become obese and develop diabetes. It seems that BDNF acts on neuronal circuits throughout the brain in ways that promote brain and body health. As described in previous chapters of this book, BDNF improves learning and memory, has antianxiety and antidepressant effects, and improves cardiovascular function. BDNF also protects us against becoming overweight and obese, and the best ways to increase BDNF levels in the brain are exercise and intermittent fasting.

AN ASS-BACKWARDS HEALTH CARE SYSTEM

Rising health care costs have been a major burden on the US economy for decades. The entire problem boils down to the fact that pharmaceutical companies, physicians, nurses, hospitals, and clinics benefit from sick people. They want to continue the processes by which they make money and have jobs. More sick people mean more dollars. The system benefits most from people with chronic illnesses that require frequent visits to specialists (cardiologists, endocrinologists, orthopedic surgeons, etc.) and daily dosing with one or more drugs. There is little incentive anywhere in the system to aggressively pursue approaches for disease prevention and risk reduction (figure 8.1). This is sad. Many recent books delve in detail into the health care crisis and its underpinnings, and readers are urged to read such books.

Ideally, the most good for the most people would be accomplished by a health care system that fosters optimal health for children from an early age. Vaccination to instill immunity against deadly contagious diseases is an excellent example of preemptive cost-reducing health care. But except in the case of an emergent crisis such as that caused by COVID-19, pharmaceutical companies usually have little interest in developing and marketing vaccines because vaccines are not very profitable in that they require only one or a few doses. Drug companies instead thrive on drugs that must be taken daily over the long term because such drugs only lessen symptoms without curing the disease. This unfortunate situation has led to the discontinuation of some vaccines. When my family and I moved to Maryland in 2000, there was a Lyme disease vaccine available. The deer ticks that carry the Lyme bacteria are prevalent on the East Coast, and so we decided to get the vaccine. However, most people were unaware that there was a Lyme disease vaccine, and so the company that produced the vaccine had to discontinue its production.

What key obstacles are hindering the incorporation of lifestyles that include intermittent fasting eating patterns into a large swath of the population?

- Health care professionals themselves are often overweight and sedentary, which can inhibit their raising the issue with their patients.
- There is a general lack of understanding of intermittent fasting among health care professionals.
- Medical school training does not include information on intermittent fasting and exercise and de-emphasizes lifestyle medicine.
- Poor-quality diet research has led to confusion among the lay public.
- Medical professionals, governments, and many individuals believe that it is too hard to change behaviors.

- Nutritional scientists need to work with behavioral scientists to implement findings in the population.

I do not envision a health care system founded on for-profit companies as fostering the best cradle-to-grave health on a society-wide basis. There is now much interest in a universal health care system, which, if done properly, would be focused on disease prevention and risk reduction. There seems to be no way around a major coordinated effort by federal, state, and local agencies, which would by necessity require a large investment of taxpayers' dollars. I envision several major fronts on such a battle against poor health.

First, the dietary factors that have clearly contributed to the obesity epidemic—sugar- and saturated-fat-laden processed foods—should be eliminated from the food supply. Although these obesogenic foods are perhaps not quite as bad as tobacco, it is my opinion that they are wreaking havoc on people's health.

Second, incentives for diversification in agricultural crops should be provided to farmers, with a plan for making a wide range of healthy vegetables and fruits available to all people at affordable prices.

Third, drug company salespersons should not be allowed to communicate with physicians, and drug advertisements should be restricted or banned.

Fourth, local metabolic health care centers should be established, and their programs coordinated by state and federal agencies. Such metabolic health care centers would include both outpatient and inpatient components, such as long-term counseling, that emphasize education and enable patients to change their eating and exercise habits over a period of several months.

Medical schools should be required to teach students the science of exercise and intermittent fasting and how to implement lifestyle-medicine prescriptions as well as to follow up in a manner that maximizes the probability of success.

A new class of physicians who are experts on metabolic health and lifestyle medicine should be established, with incentives for medical students and existing physicians to become members of this class.

The NIH should establish a Lifestyle Medicine Institute, its mission being to improve metabolic health through basic and clinical research, public outreach, and researching the influences of the processed-food and drug industries on people's behaviors.

Finally, parents should be provided information on metabolic health—why it is important and the consequences of poor metabolic health for the bodies and brains of their children.

Some of these initiatives will likely seem extreme to many readers but in my view they merit serious consideration for the health of both current and future generations.

A QUICK FIX?

There has been considerable interest among researchers in the fields of aging, obesity, and diabetes in finding chemicals that "mimic" the beneficial effects of fasting and exercise on the body and brain. Thus far, a few chemicals have been found that can elicit one or several but not all of the effects of intermittent fasting and exercise. In this section, I describe several such chemicals and provide examples of their effects on laboratory animals.

A glucose molecule called "2-deoxyglucose" (2DG) lacks an oxygen atom on one of its carbon atoms. Cells are able to take up 2DG, but, unlike glucose, 2DG cannot be used by the cells to produce ATP. Although the enzyme hexokinase is able to initiate the series of biochemical actions that produce ATP from glucose, it is unable to do the same with 2DG. Instead, 2DG prevents glucose from being acted on by hexokinase, and so impairs the cells' ability to use any glucose present to produce ATP. In effect, 2DG causes cellular calorie restriction such that even though an animal or human is consuming food and increasing glucose levels in their blood, cells in the body and brain themselves are unable to use the glucose. The body adapts to the reduced glucose availability by increasing ketone production.

In the late 1990s and early 2000s, several postdocs and graduate students in my laboratory tested the hypothesis that administration of 2DG to mice or rats would induce a mild cellular stress response and thereby protect neurons against dysfunction and degeneration in experimental models of several major neurological disorders. Wenzhen Duan found that a once-daily treatment of mice with 2DG for two weeks reduced the degeneration of dopamine neurons in the substantia nigra and alleviated motor deficits in a model of Parkinson's disease. Jaewon Lee found that a similar regimen of 2DG treatment prevented degeneration of hippocampal neurons in a mouse model of epilepsy, and Zaifang Yu showed that 2DG treatment was beneficial in a rat model of stroke. All three studies provided evidence that 2DG elicits a mild stress response in neurons, which then produce several proteins that help the neurons resist more severe stress. However, in the long-run, ingestion of 2DG may result in serious side effects. For example, when Donald Ingram was a scientist at the National Institute on Aging, he and his coworkers found that chronic ingestion of 2DG causes damage to the heart and increases the mortality of rats. Perhaps a lower dose of 2DG might avoid these side effects, but that remains to be determined. It is also possible that ingesting 2DG intermittently—every other day, for example—would improve health in a manner similar to intermittent fasting. This would allow recovery days between the 2DG challenge days.

Easter Island in the southeast Pacific Ocean is known mostly for its many large monumental stone statues made by the Rapa Nui people. However, many scientists know of Easter Island for another reason—it is where a very important drug called "rapamycin" was discovered in the bacteria species *Streptomyces hygroscopicus*. Rapamycin is used in high doses to suppress the immune system in ways that prevent the rejection of transplanted organs, and it is also used to coat coronary stents in people afflicted by coronary heart disease. In addition, by starving cancer cells of glucose, rapamycin can enhance the killing of such cells by chemotherapies. But it turns out that low doses of rapamycin can also counteract aging and extend the lifespan of mice. Rapamycin is a very specific inhibitor of mTOR, and it can stimulate autophagy and increase cellular stress resistance. These effects on mTOR are similar to those that occur in response to fasting. Neuroscientists found that rapamycin can counteract the neurodegenerative disease processes in animal models of Parkinson's and Alzheimer's diseases. There will likely be clinical trials of rapamycin and similar drugs that inhibit mTOR in neurodegenerative disorders, although concerns about risks related to immune system suppression remain.

The chemical 2,4-dinitrophenol (DNP) has a dark past but may have a bright future. An article published in 1933 reported that when overweight people were given one nontoxic dose of DNP, their metabolic rate increased, and they lost weight. Although the FDA had not approved this drug for use in humans, more than 100,000 overweight Americans took it. Among those who took the recommended dose, none had serious adverse events. However, two people took ten times the recommended dose and as a consequence died. The FDA then banned sales of DNP to the general public. Research into the effects of DNP on cellular energy metabolism then waned until about 20 years ago, when researchers discovered that DNP exerts its biological effects by acting in the mitochondria, where it causes protons (hydrogen ions) to leak across the membrane. This leaking "uncouples" electron transport, causing the mitochondria to produce heat instead of ATP. In this way, DNP causes the "burning" of calories. It turns out that there is a protein present in brown fat cells called "uncoupling protein 1" (UCP1) that, like DNP, causes a protein leak and heat production. Many mammals that live in cold climates have a lot of brown fat to keep their body warm. When it is cold outside, UCP1 kicks into gear.

By causing moderate levels of mitochondrial uncoupling, low doses of DNP impose a mild stress on cells. When Dong Liu was in my laboratory, she discovered that neurons also express an uncoupling protein normally at low levels but this is kickstarted during fasting. Liu then showed that, like fasting, DNP can cause neurons to go into a stress-resistance mode. When cultured neurons were exposed to low doses of DNP, autophagy was increased and mTOR was decreased. In addition, DNP treatment stimulated the production of the neurotrophic factor BDNF in neurons. Other studies have shown that BDNF is critical for learning and memory and helps neurons resist stress. More recently, Yuki Kishomoto found that low doses of DNP can protect neurons against degeneration and improve functional outcome in animal models of Parkinson's disease. Whether DNP will prove beneficial for humans with these brain disorders remains to be determined. Because high doses of DNP are toxic and can even cause death, any application to humans must be done with caution.

Glucagon-like peptide 1 (GLP1) is a hormone produced by cells in the intestines. It is released into the blood in response to feeding and then acts on cells in the liver and muscles to improve their sensitivity to insulin. But GLP1 stays in the blood for only a very short time (around 2 minutes) because it is chewed up by the enzyme DPPIV. Josephine Egan, Nigel Greig, Máire Doyle, and colleagues at the National Institute on Aging modified the amino sequence of GLP1 so that it would be resistant to being cleaved by DPPIV. They developed a peptide similar to GLP1 called "exenatide" and showed that it was remarkably effective in improving insulin sensitivity and reversing diabetes in animals and humans. Exenatide is now prescribed widely for patients with diabetes. With my help, Greig and scientists in his and my laboratories performed experiments aimed at determining if neurons respond to GLP1 and whether exenatide might have beneficial effects on the neurons. We first found that exenatide protects cultured neurons against excitotoxicity and that it afforded a therapeutic benefit to mice in experimental models of Parkinson's disease and stroke. Together with Greig, we also found that exenatide was beneficial in mouse models of Alzheimer's and Huntington's diseases. Greig went on to collaborate with Tim Foltyne at University College London in a double-blind, placebo-controlled trial of exenatide in Parkinson's disease patients and found that it significantly improved their motor function during a one-year treatment period. This finding is very exciting, and if larger studies also show a benefit, then exenatide may become a new treatment that slows the progression of Parkinson's disease.

Metformin is a drug widely prescribed for type 2 diabetes. It is derived from a chemical in a species of French lilac, a medicinal plant. Metformin is effective in reducing glucose levels and may elicit additional effects that are similar to intermittent fasting. This hypoglycemic effect of metformin may result from stimulation of GLP1 release from cells in the gut. In addition, metformin may directly protect cells against stress in a manner similar to fasting. Animal studies have shown that metformin can counteract aging and protect against cancers. As with fasting and exercise, metformin inhibits mTOR and stimulates autophagy, thereby preventing the abnormal accumulation of proteins such as amyloid in cells.

If you enter the abbreviation NAD on an internet search engine, top hits will include advertisements for supplements that boost levels of nicotinamide adenine dinucleotide (NAD). NAD is present in all cells, where it plays critical roles in oxidation-reduction reactions (NADH is the oxidized form of NAD, and NAD^+ is the reduced form) and the release of energy from glucose and fats. In addition, NAD^+ is necessary for the function of enzymes involved in energy metabolism, regulation of gene expression, DNA repair, and mitochondrial function. Three precursors of NAD^+ that can be taken as dietary supplements are niacin, nicotinamide, and nicotinamide riboside. NAD levels decline steadily in most cells during aging, which impairs energy metabolism and exacerbates disease processes. Both exercise and intermittent fasting can boost NAD^+ levels in cells. NAD^+-boosting supplements have been reported to be beneficial in animal models of hearing loss, cognitive decline, cardiovascular disease, diabetes, obesity, and some cancers. We found that nicotinamide and nicotinamide riboside can slow cognitive decline and lessen amyloid pathology in mouse models of Alzheimer's disease. Clinical trials of nicotinamide riboside are under way to determine whether it is beneficial for a wide range of diseases.

Some of these approaches to mimicking the health-promoting effects of intermittent fasting with dietary supplements or drugs may prove useful for some people who are unwilling to incorporate intermittent fasting and exercise into their lives. In my opinion, however, this choice could put these people on the slippery slope toward worse health. I am skeptical that any such chemicals will enable someone to over-eat, not exercise, and still remain healthy. Importantly, "quick fix" ingestion of chemicals can have serious side effects. Adaptive responses to intermittent fasting and exercise are ancient, complex, and pervasive in cells throughout the body and brain. They involve highly coordinated physiological processes that cannot be engaged by a single chemical. Good health requires that we tap into the biological processes that through millions of years of evolution have made us strong, intelligent, and resilient. Intermittent fasting, exercise, and regular intellectual challenges are the way to go. No pill can substitute for them. Challenge your body and brain, and they will pay you back—big time!

9 BON VOYAGE

A gratifying aspect of my research on intermittent fasting is that some family members and many of my friends and fellow scientists have adopted an intermittent fasting eating pattern and have experienced improvements in their health. Every now and then during the first ten years that my laboratory had been studying the effects of intermittent fasting on the brains and bodies of rats and mice, I would tell my wife, Joanne, about the quite profound beneficial effects we were observing. She would mostly just say that the findings were interesting but that it would be hard for most people to go for several days each week with little or no food or to eat all their food during a 6- or 8-hour period every day. Joanne had a subscription to *Woman's World* magazine, and shortly after the results of our 5:2 intermittent fasting study with Michelle Harvie were published in 2011, a *Woman's World* cover story focused on the study, which prompted Joanne to try the 5:2 intermittent fasting. More recently she switched to daily time-restricted eating, which both she and I find easy and effective in improving general health.

When we published the results of the 5:2 intermittent fasting study, they received a modest amount of press. As it happened, Michael Mosley, who is a physician by training and a producer at the BBC, caught wind of the findings. He produced a documentary on intermittent fasting for the BBC. He came to the United States, where he interviewed me, Valter Longo at the University of Southern California, and Krista Varaday at the University of Illinois at Chicago. As part of the documentary, Mosley tried 5:2 intermittent fasting and documented its beneficial effects on his body weight and glucose regulation. Soon after the documentary aired on the BBC in 2012, intermittent fasting became very popular in the United Kingdom. The program later aired on PBS, and within the next couple of years the term *intermittent fasting* went viral on the internet.

Numerous lab members and scientists I know at the NIH, Johns Hopkins University, and elsewhere have switched their eating pattern to intermittent fasting. For those who were overweight, their weight loss month by month after initiating intermittent fasting was obvious. Less obvious but evident when I talked with them was that their mood and ability to sustain a high level of concentration improved. In many instances, people who have adopted intermittent fasting have documented improvements in a range of clinical indicators of health, including reduced glucose, A1c, and LDL cholesterol levels. And beyond people you and I know, many famous people have incorporated intermittent fasting into their lifestyles: the actors Benedict Cumberbatch (Sherlock Holmes), Nicole Kidman, Hugh Jackman (Wolverine), Ben Affleck, Jennifer Aniston, and Miranda Kerr as well as the singers Beyoncé, Selena

Gomez, and Jennifer Lopez. They do this not only to maintain a healthy body weight but also to improve their mental clarity.

Monthly Google searches for *intermittent fasting* have risen to more than 1 million. In 2019, intermittent fasting was more widely searched than any other eating plan—Mediterranean, Keto, Noom, and so on. Top hits when searching *intermittent fasting* are "Beginner's Guides" to the eating pattern. All those websites provide pretty much the same simple description of the most common intermittent fasting regimens (5:2 and daily 16- to 18-hour fasts). These materials are alright, but finding the original scientific studies or even review articles by experts on intermittent fasting requires more digging.

A most interesting development from my perspective as a neuroscientist and someone who is now working to develop strategies that facilitate people's switching to and maintaining an intermittent fasting eating pattern is the widespread adoption of intermittent fasting by employees at some major companies in Silicon Valley and elsewhere. Having the same eating pattern as your coworkers makes intermittent fasting more enjoyable. Instead of having a "power breakfast," the group has a "power morning fast," during which plain coffee and tea are fine. Large Facebook and Slack channel intermittent fasting groups provide forums for participants to relate their own experiences, provide tips, and discuss the latest research. A common theme of the conversations in these social networks is that intermittent fasting not only improves general health and mental clarity but also enhances creativity and productivity.

Intermittent fasting can save considerable amounts of time and money. Many people, my wife and myself included, have adopted an eating pattern in which most days we skip breakfast and then eat during an 8-hour time window. Multiple moderate-size "meals" during the feeding period provide sufficient calories for weight maintenance in those of us who are already at a healthy BMI of 18–24. I suggest eating what is generally considered to be a healthy balanced diet that includes vegetables, beans, fruit, nuts, whole grains (oats and wheat), Greek yogurt, and some meat (usually fish or chicken).

In this final chapter, I provide practical advice for readers who would like to implement and optimize an intermittent fasting eating pattern.

THE "MAGIC MONTH" AND PRESCRIPTIONS FOR INTERMITTENT FASTING

It is often psychologically easier for most people to adopt an intermittent fasting eating pattern than to count the calories of every meal every day. Intermittent fasting largely eliminates obsessive contemplation of how much to eat at every meal. You focus instead on consuming no or very few calories (typically less than 600 calories) one or two days each week or, alternatively, limiting the time period you consume food and caloric beverages to 6–8 hours most days. Such intermittent fasting eating patterns can be readily incorporated into daily and weekly routines.

As described previously in this book, it takes a minimum of two to four weeks for someone who has been eating three meals plus snacks daily to become adapted to an intermittent fasting eating pattern (e.g., the 5:2 or daily 16- to 18-hour intermittent fasting). In the first week or two, some of you will be hungry and irritable during the fasting periods. Some of you may also have mild headaches and notice a reduced ability to concentrate during the fasting periods. But you can look forward to the disappearance of those initial side effects and to experiencing improvements in cognition, mood, and "energy level" within one month and for as long as you maintain your intermittent fasting eating pattern. Adapting to intermittent fasting is in many ways similar to adapting to exercise after previously being sedentary. When you are out of shape, it takes time to get in shape. When you initially begin an exercise program, the exercise may not be pleasant. But once you get in shape, you will feel great and will even not feel so well when you are unable to exercise. It likewise also takes time to get in "intermittent fasting shape."

When establishing an exercise program, you usually begin with relatively easy workouts and then progressively increase the time and intensity of the workout. A similar approach can be applied to intermittent fasting (figure 9.1). For example, let's say your goal is to compress your eating window to 6 hours every day slowly over a 3-month period. During the first month, you can reduce your eating window to 10 hours, during the second month to 8 hours, and during the third month to 6 hours. Or maybe you would like to adopt the 5:2 intermittent fasting eating pattern. You could start by eating only 600 calories for just one day each week. Do that for a month and then add another 600-calorie day each week. These are just suggestions, and you can decide how to achieve your goal based on your daily and weekly schedules. You may also consider enlisting a friend or family member to make the switch to intermittent fasting together with you so that you can encourage each other, much as exercising with someone

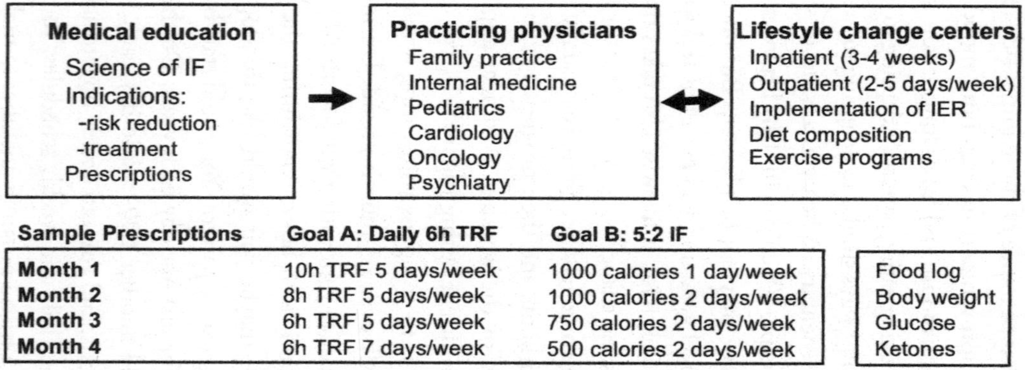

Medical education
Science of IF
Indications:
-risk reduction
-treatment
Prescriptions

Practicing physicians
Family practice
Internal medicine
Pediatrics
Cardiology
Oncology
Psychiatry

Lifestyle change centers
Inpatient (3-4 weeks)
Outpatient (2-5 days/week)
Implementation of IER
Diet composition
Exercise programs

Food log
Body weight
Glucose
Ketones

Sample Prescriptions	Goal A: Daily 6h TRF	Goal B: 5:2 IF
Month 1	10h TRF 5 days/week	1000 calories 1 day/week
Month 2	8h TRF 5 days/week	1000 calories 2 days/week
Month 3	6h TRF 5 days/week	750 calories 2 days/week
Month 4	6h TRF 7 days/week	500 calories 2 days/week

Figure 9.1

How intermittent fasting (IF) prescriptions can be incorporated into health care systems (IER=intermittent energy restriction; TRE=daily time-restricted eating). A modified version of this illustration was previously published in R. de Cabo and M. P. Mattson, "Impact of Intermittent Fasting on Health, Aging, and Disease," *New England Journal of Medicine* 381, no. 26 (Dec. 2019): 2541–2551.

else can make it easier, at least psychologically. Establishing intermittent fasting groups at the workplace can help avoid the awkward situation of all of your friends eating at a time when you are fasting..

Physicians should know how to prescribe intermittent fasting, how to help the patient transition to the new eating pattern, and how to monitor their patient's progress over periods of months and years (figure 9.1). The physician would discuss the plan and keep in touch with the patient by text messaging or internet portals on a daily basis during the first week and on a weekly basis during the first two months. After that period, the patient would return to the doctor's office for a follow-up visit, at which time her body weight, pulse rate, and blood pressure would be measured, and blood would be drawn for measurements of glucose and A1c levels. Ideally, blood would be drawn at the end of the person's daily or twice weekly fasting period, and ketone levels would be measured. Assuming the person is indeed following her intermittent fasting prescription, then by the end of the two-month period she would be expected to have lost some weight and to have lower blood pressure, heart rate, glucose, and A1c levels compared to what they were before she began intermittent fasting. The patient would then be seen every six months for repeat evaluations and consultation.

An increasing number of physicians are becoming aware of intermittent fasting as a complement to exercise and a healthy diet. Its ability to help people lose weight and reduce their risk for many diseases is prompting some physicians to recommend intermittent fasting to overweight patients and those with insulin resistance or an unhealthy blood lipid profile.

If the results of clinical trials of intermittent fasting in patients with diabetes, cancers, cardiovascular disease, and inflammatory diseases show benefits of intermittent fasting, then we can expect that intermittent fasting will be prescribed for patients with these diseases. Stay tuned!

INTERMITTENT FASTING HACKS

How might you enhance the health benefits of intermittent fasting? There are several ways. One approach is to exercise in the fasted state. For example, if you are on an intermittent fasting regimen in which you skip breakfast, you can then exercise at noon and eat the first food of your day after working out. The exercise will give a further boost to ketone production, with all the benefits of ketones being accentuated. As we discovered in our animal studies, exercise during fasting also enhances the production of BDNF and the formation of new synapses in the brain. Because intermittent fasting or exercise can on its own improve learning and memory, the combination of the two can help optimize cognitive performance. Because fasting or exercise on its own stimulates autophagy and mitochondrial biogenesis, the combination will enhance the clearance of molecular garbage and enhance cellular energy metabolism. Many bodybuilders do not eat breakfast, do their weight training midday, and then eat afterward. They have found that this eating pattern enables them to build muscle mass while reducing fat mass and thereby accentuate their muscle appearance.

There are now companies that promote and sell fatty acid precursors of ketones, so-called medium-chain triglycerides, or MCTs. Their pitch is that if you consume MCTs during the fasting period, it will give a further boost to the elevated ketones produced from your own fats. This claim seems reasonable, and there have been a few published studies in which MCTs improved performance in patients with mild cognitive impairment or Alzheimer's disease. If and to what extent MCTs boost cognition in people who are already functioning at a high level remains to be determined. The ketone ester that was the focus of chapter 5 can elevate circulating ketone levels to a much higher level than MCTs and might therefore be expected to be better than MCTs in enhancing cognition. It would be of considerable interest to compare the ketone ester and MCTs head to head in a controlled study of their use by cognitively normal people as well as by those with cognitive impairment. Similarly, comparisons of the effects on endurance of intermittent fasting alone or in combination with MCTs or the ketone ester may lead to new approaches to enhancing athletic performance.

Another intermittent fasting hack is to consume moderate amounts of caffeine during the fasting period. Caffeine increases alertness and can also improve learning and memory somewhat. Coffee or tea during the fasting period might enhance the beneficial effects of fasting on cognition. We do know that caffeine inhibits the receptors for a neurotransmitter called adenosine and that this explains caffeine's ability to promote sustained cognitive performance. Chapter 6 described how many bitter-tasting chemicals in vegetables, fruits, and other plants can stimulate beneficial mild stress responses in cells. Although it remains to be established, it seems likely that phytochemicals such as curcumin, sulforaphane, and resveratrol can enhance many of the beneficial effects of intermittent fasting on health. However, exercise and intermittent fasting improve health by complex and highly coordinated effects on cells and organs, whereas the effects of phytochemicals are much more limited.

THE UNCERTAIN FUTURE OF THE BRAIN

What will the brain look like in the next century? Will it be bigger or smaller? Will it be less or more prone to mental illness and diseases of aging? If the recent epidemic of obesity and sedentariness continues unabated, I fear that the performance and lifelong health of the brains of future generations will suffer. In this book, you have learned about the importance of the ability to overcome food scarcity in the evolutionary history of the human brain. You have learned how intermittent fasting can tap evolutionarily ancient responses of cells in the brain and body in ways that optimize their performance and resistance to chronic diseases. As part of a lifestyle that includes regular exercise and a healthy diet, intermittent fasting could facilitate a reversal of many Americans' declining health. It has also become clear that poor brain health can be inflicted on future generations by epigenetic mechanisms. That is to say, the metabolic state of parents influences the health trajectories of their children. The evidence that children born to mothers with obesity or diabetes or both are more likely to develop mental illnesses such as autism, anxiety disorders, or depression is particularly troubling. In this final section of the book, I ruminate on the future of the human brain in light of the current state of understanding of how intermittent fasting affects brain health throughout the life course.

Being metabolically unhealthy—overweight, obese, insulin resistant, and abdominally fat—greatly increases one's risk for stroke. Strokes most commonly occur in people older than 65, and the longer one is metabolically unhealthy during one's lifetime the more likely one is to have a stroke. But stroke is not the only danger that poor metabolic health poses for the brain as one grows older. There is now considerable evidence that being metabolically unhealthy can increase one's risk for Alzheimer's disease and Parkinson's disease. In the United States, the Baby Boomers are now into or entering the danger-zone ages for these major neurodegenerative disorders, and many of these people have led metabolically unhealthy lifestyles. If the epidemic of poor metabolic health continues, the country is going to suffer a major blow to families, with major personal, social, and economic costs.

If you are reading this book, you should assume that there is a good chance that you will become afflicted with Alzheimer's disease, Parkinson's disease, or stroke as you transit into your seventh and eighth decades of life. The lifestyle changes I have discussed here can reduce your risk: regular exercise, moderation in energy intake, intermittent fasting, and intellectual challenges for the mind. As I worked to understand what goes wrong in the brain in Alzheimer's and Parkinson's diseases, it became obvious that there will be no miracle drug that reverses the havoc that has already occurred in neuronal networks in patients with these diseases. Accordingly, it became clear that lifelong risk reduction should be a priority.

One environmental factor that can help neurons maintain their structure and function is intermittent fasting. In contrast to a drug that acts on one molecular target, intermittent fasting sets in motion complex and highly integrated changes in neurons that include the bolstering of mitochondrial function and stress resistance; enhanced removal and recycling of molecular garbage; production of neurotrophic factors; and changes in the activity of and synaptic connections between neurons that improve brain function and resilience. These and other neuronal network adaptations to intermittent fasting have been honed over millions of years of evolution. I do not expect that intermittent fasting will cure someone who has already been diagnosed with Alzheimer's or Parkinson's disease, but it may improve symptoms and slow disease progression. When initiated in midlife, intermittent fasting may prevent or delay the age of onset of these brain disorders.

In addition to intermittent fasting, exercise, and intellectual challenges, other potential ways to improve brain function and resilience are being tested. In some cases, many people are self-experimenting with such approaches. They include application of low-power direct-current stimulation (DCS) of the brain and microdosing with hallucinogenic chemicals such as psilocybin (found in "magic mushrooms") and LSD. There is emerging evidence from clinical trials that DCS can enhance cognition, but the ideal stimulation parameters (current intensity, duration of stimulation, and frequency of stimulation) remain to be established. There is now strong evidence that psilocybin can alleviate depression. To my knowledge, there have been no controlled clinical trials to determine whether microdosing can improve cognition. Such trials should be done.

One promising approach for improving brain function and treating certain psychiatric disorders and drug or alcohol addiction is transcranial magnetic stimulation (TMS). Many studies have shown the benefits of TMS in improving cognition in older adults and in treating depression and drug or alcohol addiction. These results are exciting because TMS is safe, and in many cases only one or a few treatments result in fairly long-lasting beneficial effects. Whether this technology can be commercialized at a reasonable price and standardized for use by the general public remains to be determined.

In the 1990s, neuroscientists established that at least two brain regions have stem cells that are capable of producing new neurons that can integrate into neuronal networks. This raised the possibility that stem cells could be transplanted into the brains of patients with Alzheimer's or Parkinson's disease and so replace the neurons that had died. There was excitement about the potential of stem cell therapies, but this approach has proven problematic, with many technical hurdles. Animal studies suggested this approach might work, but clinical trials in patients with Parkinson's disease failed. Most of the stem cells that were transplanted died. Then the Japanese scientist Shinya Yamanaka developed a method by which differentiated cells such as fibroblasts from the skin can be transformed into so-called pluripotent stem cells, which are the kind of stem cell present in the very earliest stages of development of an embryo. The pluripotent stem cells have the ability to become any type of cell in the body and brain, and which cell type they become depends on the signals they receive from the cells that surround them. Very recent findings suggest that it may be possible to cause one cell type to transform into another cell type by expressing only a few genes. We can therefore envision a time in the distant future in which someone with Alzheimer's disease can be treated with a gene therapy that causes glial cells to become neurons that replace the neurons that have died.

The interfacing of the human brain with machines is moving forward rapidly. In the future, you may be able to sit on your front porch and mow your lawn via mind control of a mower with no seat, foot pedals, or steering wheel. Of course, there are more important applications of such technologies. One exciting medical application is for people who have had a limb amputated. Working with engineers, neuroscientists and neurologists have shown that it is possible for these patients to control an artificial limb with their mind. But artificial intelligence and computer–brain interface devices are unlikely to be of great benefit to a patient with Alzheimer's disease who has already lost millions of neurons in neuronal networks that are essential for learning and memory. It is therefore important that people who would like to avoid Alzheimer's disease not succumb to the tendency to become sedentary and intellectually disengaged.

What are the prospects for evolution of our brains in the coming millennia? As noted in chapter 1, there is evidence that the overall brain size of *Homo sapiens* has decreased during the past approximately 10,000 years, which corresponds to the period of the Agricultural Revolution. Will continued specialization of occupations result in shrinkage of some brain regions and enlargement of others? Will increasingly sedentary and overindulgent lifestyles as well as reliance on drugs send the human brain down a rabbit hole? Will *Homo sapiens* split into two species with different brains—those diminished as a consequence of transgenerational metabolic morbidity and those bolstered by healthy lifestyles? Such a scenario is consistent with an evolutionary history in which oversize Neanderthals were superseded by more svelte humans. It is also consistent with evidence described in chapter 7 showing that a sedentary, overindulgent lifestyle adversely affects not only potential parents' brains but also their offspring's brains.

I hope that the information in this book encourages readers to consider how their lifestyle affects their brain and body—their performance and susceptibility to disease. Humans will continue to age and die of diseases for many, many generations into the future. We can all do our part to reduce the burden of chronic diseases by taking responsibility for our own health and by helping others do the same.

Further Reading

CHAPTER 1

Clayton, N. S. "Ways of Thinking: From Crows to Children and Back Again." *Quarterly Journal of Experimental Psychology (Hove)* 68, no. 2 (2015): 209–241.

Dewey, Edward. *The No-Breakfast Plan and the Fasting Cure*. New York: L. N. Fowler, 1900.

Grundler, F., R. Mesnage, N. Goutzourelas, F. Tekos, S. Makri, M. Brack, D. Kouretas, and F. Wilhelmi de Toledo. "Interplay between Oxidative Damage, the Redox Status, and Metabolic Biomarkers during Long-Term Fasting." *Food and Chemical Toxicology* 145 (2020): 111701.

Hazzard, Linda. *Fasting for the Cure of Disease*. New York: Physical Culture Publishing Company, 1908.

Mattson, M. P. "Evolutionary Aspects of Human Exercise—Born to Run Purposefully." *Ageing Research Reviews* 11 (2012): 347–352.

Mattson, M. P. "An Evolutionary Perspective on Why Food Overconsumption Impairs Cognition." *Trends in Cognitive Sciences* 23, no. 3 (2019): 200–212.

Mattson, M. P. "Lifelong Brain Health Is a Lifelong Challenge: From Evolutionary Principles to Empirical Evidence." *Ageing Research Reviews* 20 (2015): 37–45.

Morriss-Kay, G. M. "The Evolution of Human Artistic Creativity." *Journal of Anatomy* 216, no. 2 (2010): 158–176.

Norenzayan, A., and A. F. Shariff. "The Origin and Evolution of Religious Prosociality." *Science* 322, no. 5898 (Oct. 3, 2008): 58–62.

Pearson, J. M., K. K. Watson, and M. L. Platt. "Decision Making: The Neuroethological Turn." *Neuron* 82, no. 5 (2014): 950–965.

Peoples, H. C., and F. W. Marlowe. "Subsistence and the Evolution of Religion." *Human Nature* 23, no. 3 (2012): 253–269.

Sinclair, Upton. *The Fasting Cure.* New York: Mitchell Kennerley, 1911.

CHAPTER 2

Anderson, R. M., and R. Weindruch. "Metabolic Reprogramming, Caloric Restriction, and Aging." *Trends in Endocrinology and Metabolism* 21 (2010): 134–141.

Anson, R. M., Z. Guo, R. de Cabo, T. Iyun, M. Rios, A. Hagepanos, D. K. Ingram, et al. "Intermittent Fasting Dissociates Beneficial Effects of Dietary Restriction on Glucose Metabolism and Neuronal Resistance to Injury from Calorie Intake." *Proceedings of the National Academy of Sciences* 100, no. 10 (2003): 6216–6220.

Anton, S. D., K. Moehl, W. T. Donahoo, K. Marosi, S. A. Lee, A. G. Mainous III, C. Leeuwenburgh, and M. P. Mattson. "Flipping the Metabolic Switch: Understanding and Applying the Health Benefits of Fasting." *Obesity* 26 (2018): 254–268.

Baker, D. J., T. Wijshake, T. Tchkonia, N. K. LeBrasseur, B. G. Childs, B. van de Sluis, J. L. Kirkland, et al. "Clearance of p16Ink4a-positive Senescent Cells Delays Ageing-Associated Disorders." *Nature* 479, no. 7372 (2011): 232–236.

Carlson, A., and F. Hoelzel. "Apparent Prolongation of the Life Span of Rats by Intermittent Fasting." *Journal of Nutrition* 31 (1946): 363–375.

Cleynen, I., and S. Vermeire. "Paradoxical Inflammation Induced by Anti-TNF Agents in Patients with IBD." *Nature Reviews Gastroenterology and Hepatology* 9, no. 9 (2012): 496–503.

Di Francesco, A., C. Di Germanio, M. Bernier, and R. de Cabo. "A Time to Fast." *Science* 362, no. 6416 (2018): 770–775.

Donati, A., G. Recchia, G. Cavallini, and E. J. Bergamini. "Effect of Aging and Anti-Aging Caloric Restriction on the Endocrine Regulation of Rat Liver Autophagy." *Journal of Gerontology A* 63, no. 6 (2008): 550–555.

Donato, A. J., A. E. Walker, K. A. Magerko, R. C. Bramwell, A. D. Black, G. D. Henson, B. R. Lawson, et al. "Life-Long Caloric Restriction Reduces Oxidative

Stress and Preserves Nitric Oxide Bioavailability and Function in Arteries of Old Mice." *Aging Cell* 12, no. 5 (2013): 772–783.

Goodrick, C. L., D. K. Ingram, M. A. Reynolds, J. R. Freeman, and N. L. Cider. "Differential Effects of Intermittent Feeding and Voluntary Exercise on Body Weight and Lifespan in Adult Rats." *Journal of Gerontology* 38 (1983): 36–45.

Goodrick, C. L., D. K. Ingram, M. A. Reynolds, J. R. Freeman, and N. Cider. "Effects of Intermittent Feeding upon Body Weight and Lifespan in Inbred Mice: Interaction of Genotype and Age." *Mechanisms of Ageing and Development* 55, no. 1 (1990): 69–87.

Ingram, D. K., E. D. London, and C. L. Goodrick. "Age and Neurochemical Correlates of Radial Maze Performance in Rats." *Neurobiology of Aging* 2 (1981): 41–47.

Jacob, F. *The Possible and the Actual*. New York. Pantheon Books, 1982.

Kjeldsen-Kragh, J., M. Haugen, C. F. Borchgrevink, E. Laerum, M. Eek, P. Mowinkel, K. Hovi, et al. "Controlled Trial of Fasting and One-Year Vegetarian Diet in Rheumatoid Arthritis." *Lancet* 338, no. 8772 (1991): 899–902.

Knapp, L. T., and E. Klann. "Potentiation of Hippocampal Synaptic Transmission by Superoxide Requires the Oxidative Activation of Protein Kinase C." *Journal of Neuroscience* 22, no. 3 (2002): 674–683.

Li, G., C. Xie, S. Lu, R. G. Nichols, Y. Tian, L. Li, D. Patel, et al. "Intermittent Fasting Promotes White Adipose Browning and Decreases Obesity by Shaping the Gut Microbiota." *Cell Metabolism* 26, no. 4 (2017): 672–685.

Longo, V. D., and M. P. Mattson. "Fasting: Molecular Mechanisms and Clinical Applications." *Cell Metabolism* 19 (2014): 181–192.

Lu, T., Y. Pan, S.-Y. Kao, C. Li, I. Kohane, J. Chan, and B. A. Yankner. "Gene Regulation and DNA Damage in the Ageing Human Brain." *Nature* 429, no. 6994 (2004): 883–891.

Mattison, J. A., R. J. Colman, T. M. Beasley, D. B. Allison, J. W. Kemnitz, G. S. Roth, D. K. Ingram, et al. "Caloric Restriction Improves Health and Survival of Rhesus Monkeys." *Nature Communications* 8 (2017): 14063.

Mattson, M. P., and T. V. Arumugam. "Hallmarks of Brain Aging: Adaptive and Pathological Modification by Metabolic States." *Cell Metabolism* 27 (2018): 1176–1199.

Mattson, M. P., K. Moehl, N. Ghena, M. Schmaedick, and A. Cheng. "Intermittent Metabolic Switching, Neuroplasticity, and Brain Health." *Nature Reviews Neuroscience* 19 (2018): 63–80.

Meydani, S. N., S. K. Das, C. F. Pieper, M. R. Lewis, S. Klein, V. D. Dixit, A. K. Gupta, et al. "Long-Term Moderate Calorie Restriction Inhibits Inflammation without Impairing Cell-Mediated Immunity: A Randomized Controlled Trial in Non-Obese Humans." *Aging* 8, no. 7 (2016): 1416–1431.

Mitchell, S. J., M. Bernier, J. A. Mattison, M. A. Aon, T. A. Kaiser, R. M. Anson, Y. Ikeno, et al. "Daily Fasting Improves Health and Survival in Male Mice Independent of Diet Composition and Calories." *Cell Metabolism* 29, no. 1 (2019): 221–228.e3.

Nixon, Ralph A. "The Role of Autophagy in Neurodegenerative Disease." *Nature Medicine* 19, no. 8 (2013): 983–997.

Panda, S. *The Circadian Code: Lose Weight, Supercharge Your Energy, and Transform Your Health from Morning to Midnight*. Kutztown, PA: Rodale Institute, 2018.

Rangan, P., I. Choi, M. Wei, G. Navarrete, E. Guen, S. Brandhorst, N. Enyati, et al. "Fasting-Mimicking Diet Modulates Microbiota and Promotes Intestinal Regeneration to Reduce Inflammatory Bowel Disease Pathology." *Cell Reports* 26, no. 10 (2019): 2704–2719.

Ravussin, E., L. M. Redman, J. Rochon, S. Krupa Das, L. Fontana, W. E. Kraus, S. Romashkan, et al. "A 2-Year Randomized Controlled Trial of Human Caloric Restriction: Feasibility and Effects on Predictors of Health Span and Longevity." *Journal of Gerontology* 70 (2015): 1097–1104.

Rubinsztein, D. C., G. Mariño, and G. Kroemer. "Autophagy and Aging." *Cell* 146, no. 5 (2011): 682–695.

Snyder, S. H., and D. S. Bredt. "Nitric Oxide as a Neuronal Messenger." *Trends in Pharmacological Sciences* 12 (1991): 125–128.

Sutton, E. F., R. Beyl, K. S. Early, W. T. Cefalu, E. Ravussin, and C. M. Peterson. "Early Time-Restricted Feeding Improves Insulin Sensitivity, Blood Pressure, and Oxidative Stress Even without Weight Loss in Men with Prediabetes." *Cell Metabolism* 27, no. 6 (2018): 1212–1221.

Weissman, L., D. G. Jo, M. M. Sorensen, N. C. de Souza-Pinto, W. R. Markesbery, M. P. Mattson, and V. A. Bohr. "Defective DNA Base Excision Repair in

Brain from Individuals with Alzeimer's Disease and Amnestic Mild Cognitive Impairment." *Nucleic Acids Research* 35, no. 16 (2007): 5545–5555.

Zhang, P., Y. Kishimoto, I. Grammatikakis, K. Gottimukkala, R. G. Cutler, S. Zhang, K. Abdelmohsen, et al. "Senolytic Therapy Alleviates Aβ-Associated Oligodendrocyte Progenitor Cell Senescence and Cognitive Deficits in an Alzeimer's Disease Model." *Nature Neuroscience* 22, no. 5 (2019): 719–728.

CHAPTER 3

Ahmet, I., R. Wan, M. P. Mattson, E. G. Lakatta, and M. Talan. "Cardioprotection by Intermittent Fasting in Rats." *Circulation* 112, no. 20 (2005): 3115–3121.

Anton, S. D., K. Moehl, W. T. Donahoo, K. Marosi, S. A. Lee, A. G. Mainous III, C. Leeuwenburgh, and M. P. Mattson. "Flipping the Metabolic Switch: Understanding and Applying the Health Benefits of Fasting." *Obesity* 26 (2018): 254–268.

Antoni, R., K. L. Johnston, A. L. Collins, and M. D. Robertson. "Intermittent v. Continuous Energy Restriction: Differential Effects on Postprandial Glucose and Lipid Metabolism Following Matched Weight Loss in Overweight/Obese Participants." *British Journal of Nutrition* 119, no. 5 (Mar. 2018): 507–516.

Arumugam, T. V., T. M. Phillips, A. Cheng, C. H. Morrell, M. P. Mattson, and R. Wan. "Age and Energy Intake Interact to Modify Cell Stress Pathways and Stroke Outcome." *Annals of Neurology* 67, no. 1 (2010): 41–52.

Carter, S., P. M. Clifton, and J. B. Keogh. "The Effects of Intermittent Compared to Continuous Energy Restriction on Glycaemic Control in Type 2 Diabetes: A Pragmatic Pilot Trial." *Diabetes Research Clinical Practice* 122 (2016): 106–112.

Chaix, A., E. N. C. Manoogian, G. C. Melkani, and S. Panda. "Time-Restricted Eating to Prevent and Manage Chronic Metabolic Diseases." *Annual Review Nutrition* 39 (Aug. 21, 2019): 291–315.

Choi, I. Y., C. Lee, and V. D. Longo. "Nutrition and Fasting Mimicking Diets in the Prevention and Treatment of Autoimmune Diseases and Immunosenescence." *Molecular and Cellular Endocrinology* 455 (2017): 4–12.

Choi, I. Y., L. Piccio, P. Childress, B. Bollman, A. Ghosh, S. Brandhorst, J. Suarez, et al. "A Diet Mimicking Fasting Promotes Regeneration and Reduces Autoimmunity and Multiple Sclerosis Symptoms." *Cell Reports* 15, no. 10 (2016): 2136–2146.

De Cabo, R., and M. P. Mattson. "Impact of Intermittent Fasting on Health, Aging, and Disease." *New England Journal of Medicine* 381, no. 26 (Dec. 2019): 2541–2551.

Di Biase, S., C. Lee, S. Brandhorst, B. Manes, R. Buono, C.-W. Cheng, M. Cacciottolo, et al. "Fasting-Mimicking Diet Reduces HO-1 to Promote T Cell–Mediated Tumor Cytotoxicity." *Cancer Cell* 30, no. 1 (2016): 136–146.

Donato, A. J., A. E. Walker, K. A. Magerko, R. C. Bramwell, A. D. Black, G. D. Henson, B. R. Lawson, et al. "Life-Long Caloric Restriction Reduces Oxidative Stress and Preserves Nitric Oxide Bioavailability and Function in Arteries of Old Mice." *Aging Cell* 12, no. 5 (2013): 772–783.

Duan, W., Z. Guo, H. Jiang, M. Ware, and M. P. Mattson. "Reversal of Behavioral and Metabolic Abnormalities, and Insulin Resistance Syndrome, by Dietary Restriction in Mice Deficient in Brain-Derived Neurotrophic Factor." *Endocrinology* 144, no. 6 (2003): 2446–2453.

Duan, W., and M. P. Mattson. "Dietary Restriction and 2-Deoxyglucose Administration Improve Behavioral Outcome and Reduce Degeneration of Dopaminergic Neurons in Models of Parkinson's Disease." *Journal of Neuroscience Research* 57, no. 2 (1999): 195–206.

Fitzgerald, K. C., D. Vizthum, B. Henry-Barron, A. Schweitzer, S. D. Cassard, E. Kossoff, A. L. Hartman, et al. "Effect of Intermittent vs. Daily Calorie Restriction on Changes in Weight and Patient-Reported Outcomes in People with Multiple Sclerosis." *Multiple Sclerosis and Related Disorders* 23 (2018): 33–39.

Griffioen, K. J., S. M. Rothman, B. Ladenheim, R. Wan, N. Vranis, E. Hutchison, E. Okun, et al. "Dietary Energy Intake Modifies Brainstem Autonomic Dysfunction Caused by Mutant α-synuclein." *Neurobiology of Aging* 34, no. 3 (2013): 928–935.

Halagappa, V. K. M., Z. Guo, M. Pearson, Y. Matsuoka, R. G. Cutler, F. M. Laferla, and M. P. Mattson. "Intermittent Fasting and Caloric Restriction Ameliorate Age-Related Behavioral Deficits in the Triple-Transgenic Mouse Model of Alzheimer's Disease." *Neurobiology of Disease* 26, no. 1 (2007): 212–220.

Harvie, M. N., M. Pegington, M. P. Mattson, J. Frystyk, B. Dillon, G. Evans, J. Cuzick, et al. "The Effects of Intermittent or Continuous Energy Restriction on Weight Loss and Metabolic Disease Risk Markers: A Randomized Trial in Young Overweight Women." *International Journal of Obesity* 35 (2011): 714–727.

Heilbronn, L. K., S. R. Smith, C. K. Martin, S. D. Anton, and E. Ravussin. "Alternate-Day Fasting in Nonobese Subjects: Effects on Body Weight, Body Composition, and Energy Metabolism." *American Journal of Clinical Nutrition* 81 (2005): 69–73.

Hu, Y., M. Zhang, Y. Chen, Y. Yang, and J. J. Zhang. "Postoperative Intermittent Fasting Prevents Hippocampal Oxidative Stress and Memory Deficits in a Rat Model of Chronic Cerebral Hypoperfusion." *European Journal of Nutrition* 58, no. 1 (2018): 423–432.

Jebeile, H., M. L. Gow, N. B. Lister, M. Mosalman Haghighi, J. Ayer, C. T. Cowell, L. A. Baur, and S. P. Garnett. "Intermittent Energy Restriction Is a Feasible, Effective, and Acceptable Intervention to Treat Adolescents with Obesity." *Journal of Nutrition* 149 (2019): 1189–1197.

Johnson, J. B., W. Summer, R. G. Cutler, B. Martin, and D. H. Hyun. "Alternate Day Calorie Restriction Improves Clinical Findings and Reduces Markers of Oxidative Stress and Inflammation in Overweight Adults with Moderate Asthma." *Free Radical Biology and Medicine* 42 (2007): 665–674.

Kishimoto, Y., W. Zhu, W. Hosada, J. M. Sen, and M. P. Mattson. "Chronic Mild Gut Inflammation Accelerates Brain Neuropathology and Motor Dysfunction in α-Synuclein Mutant Mice." *Neuromolecular Medicine* 21, no. 3 (2019): 239–249.

Kjeldsen-Kragh, J., M. Haugen, C. F. Borchgrevink, E. Laerum, M. Eek, P. Mowinkel, K. Hovi, et al. "Controlled Trial of Fasting and One-Year Vegetarian Diet in Rheumatoid Arthritis." *Lancet* 338, no. 8772 (1991): 899–902.

Kroeger, C. M., M. C. Klempel, S. Bhutani, J. F. Trepanowski, C. C. Tangney, and K. A. Varady. "Improvement in Coronary Heart Disease Risk Factors during an Intermittent Fasting/Calorie Restriction Regimen: Relationship to Adipokine Modulations." *Nutrition and Metabolism* (London) 9, no. 1 (2012): 98.

Lefevre, M., L. M. Redman, L. K. Heilbronn, J. V. Smith, C. K. Martin, J. C. Rood, F. L. Greenway, et al. "Caloric Restriction Alone and with Exercise Improves CVD Risk in Healthy Non-obese Individuals." *Atherosclerosis* 203 (2009): 206–213.

Li, G., C. Xie, S. Lu, R. G. Nichols, Y. Tian, L. Li, D. Patel, et al. "Intermittent Fasting Promotes White Adipose Browning by Shaping the Gut Microbiota." *Cell Metabolism* 26 (2017): 672–685.e4.

Longo, V. D., and M. P. Mattson. "Fasting: Molecular Mechanisms and Clinical Applications." *Cell Metabolism* 19 (2014): 181–192.

Maswood, N., J. Young, E. Tilmont, Z. Zhang, D. M. Gash, G. A. Gerhardt, R. Grondin, et al. "Caloric Restriction Increases Neurotrophic Factor Levels and Attenuates Neurochemical and Behavioral Deficits in a Primate Model of Parkinson's Disease." *Proceedings of the National Academy of Sciences* 101, no. 52 (2004): 18171–18176.

Nencioni, A., I. Caffa, S. Cortellino, and V. D. Longo. "Fasting and Cancer: Molecular Mechanisms and Clinical Application." *Nature Review Cancer* 18 (2018): 707–719.

Okoshi, K., M. D. M. Cezar, M. A. M. Polin, J. R. Paladino Jr., P. F. Martinez, S. A. Oliveira Jr., A. R. R. Lima, et al. "Influence of Intermittent Fasting on Myocardial Infarction–Induced Cardiac Remodeling." *BMC Cardiovascular Disorders* 19, no. 1 (2019): 126.

Ross, M. H., and G. Bras. "Lasting Influence of Early Caloric Restriction on Prevalence of Neoplasms in the Rat." *Journal of the National Cancer Institute* 47 (1971): 1095–1113.

Sundfor, T. M., M. Svendsen, and S. Tonstad. "Effect of Intermittent versus Continuous Energy Restriction on Weight Loss, Maintenance and Cardiometabolic Risk: A Randomized 1-Year Trial." *Nutrition, Metabolism, and Cardiovascular Diseases* 28, no. 7 (2018): 698–706.

Trepanowski, J. F., C. M. Kroeger, A. Barnosky, M. C. Klempel, S. Bhutani, and K. K. Hoddy. "Effect of Alternate-Day Fasting on Weight Loss, Weight Maintenance, and Cardioprotection among Metabolically Healthy Obese Adults: A Randomized Clinical Trial." *JAMA Internal Medicine* 177 (2017): 930–938.

Vermeij, W. P., M. E. T. Dollé, E. Reiling, D. Jaarsma, C. Payan-Gomez, C. R. Bombardieri, H. Wu, et al. "Restricted Diet Delays Accelerated Ageing and Genomic Stress in DNA-Repair-Deficient Mice." *Nature* 537, no. 7620 (2016): 427–431.

Wan, R., L. A. Weigand, R. Bateman, K. Griffioen, D. Mendelowitz, and M. P. Mattson. "Evidence That BDNF Regulates Heart Rate by a Mechanism Involving Increased Brainstem Parasympathetic Neuron Excitability." *Journal of Neurochemistry* 129 (2014): 573–580.

Wilkinson, M. J., E. N. C. Manoogian, A. Zadourian, H. Lo, S. Fakhouri, A. Shoghi, X. Wang, et al. "Ten-Hour Time-Restricted Eating Reduces Weight,

Blood Pressure, and Atherogenic Lipids in Patients with Metabolic Syndrome." *Cell Metabolism* 31, no. 1 (2020): 92–104.

Yang, J.-L., Y.-T. Lin, P.-C. Chuang, V. A. Bohr, and M. P. Mattson. "BDNF and Exercise Enhance Neuronal DNA Repair by Stimulating CREB-Mediated Production of Apurinic/Apyrimidinic Endonuclease 1." *Neuromolecular Medicine* 16, no. 1 (2014): 161–174.

Yu, Z. F., and M. P. Mattson. "Dietary Restriction and 2-Deoxyglucose Administration Reduce Focal Ischemic Brain Damage and Improve Behavioral Outcome: Evidence for a Preconditioning Mechanism." *Journal of Neuroscience Research* 57, no. 6 (1999): 830–839.

CHAPTER 4

Bruce-Keller, A. J., J. M. Salbaum, M. Luo, E. Blanchard IV, C. M. Taylor, D. A. Welsh, and H. R. Berthoud. "Obese-Type Gut Microbiota Induce Neurobehavioral Changes in the Absence of Obesity." *Biological Psychiatry* 77 (2015): 607–615.

Camandola, S., and M. P. Mattson. "Brain Metabolism in Health, Aging, and Neurodegeneration." *EMBO Journal* 36 (2017): 1474–1492.

Cheng, A., R. Wan, J. L. Yang, N. Kamimura, T. G. Son, X. Ouyang, Y. Luo, et al. "Involvement of PGC-1α in the Formation and Maintenance of Neuronal Dendritic Spines." *Nature Communications* 3 (2012): 1250.

Cheng, A., Y. Yang, Y. Zhou, C. Maharana, D. Lu, W. Peng, Y. Liu, et al. "Mitochondrial SIRT3 Mediates Adaptive Responses of Neurons to Exercise and Metabolic and Excitatory Challenges." *Cell Metabolism* 23 (2016): 128–142.

Cignarella, F., C. Cantoni, L. Ghezzi, A. Salter, Y. Dorsett, L. Chen, D. Phillips, et al. "Intermittent Fasting Confers Protection in CNS Autoimmunity by Altering the Gut Microbiota." *Cell Metabolism* 27 (2018): 1222–1235.e6.

Jones, R., P. Pabla, J. Mallinson, A. Nixon, T. Taylor, A. Bennett, and K. Tsintzas. "Two Weeks of Early Time-Restricted Feeding Improves Skeletal Muscle Insulin and Anabolic Sensitivity in Healthy Men." *American Journal of Clinical Nutrition* 112, no. 4 (July 30, 2020): 1015–1028.

Lee, J., K. B. Seroogy, and M. P. Mattson. "Dietary Restriction Enhances Neurotrophin Expression and Neurogenesis in the Hippocampus of Adult Mice." *Journal of Neurochemistry* 80, no. 3 (2002): 539–547.

Li, G., C. Xie, S. Lu, R. G. Nichols, Y. Tian, L. Li, D. Patel, et al. "Intermittent Fasting Promotes White Adipose Browning and Decreases Obesity by Shaping the Gut Microbiota." *Cell Metabolism* 26 (2017): 672–685.e4.

Liu, Y., A. Cheng, Y-J. Li, Y. Yang, Y. Kishimoto, S. Zhang, Y. Wang, et al. "SIRT3 Mediates Hippocampal Synaptic Adaptations to Intermittent Fasting and Ameliorates Deficits in APP Mutant Mice." *Nature Communications* 10, no. 1 (2019): 1886.

Mager, D. E., R. Wan, M. Brown, A. Cheng, P. Wareski, D. R. Abernethy, and M. P. Mattson. "Caloric Restriction and Intermittent Fasting Alter Spectral Measures of Heart Rate and Blood Pressure Variability in Rats." *FASEB Journal* 20 (2006): 631–637.

Marosi, K., S. W. Kim, K. Moehl, M. Scheibye-Knudsen, A. Cheng, R. Cutler, S. Camandola, and M. P. Mattson. "3-Hydroxybutyrate Regulates Energy Metabolism and Induces BDNF Expression in Cerebral Cortical Neurons." *Journal of Neurochemistry* 139 (2016): 769–781.

Marosi, K., K. Moehl, I. Navas-Enamorado, S. J. Mitchell, Y. Zhang, E. Lehrmann, M. A. Aon, et al. "Metabolic and Molecular Framework for the Enhancement of Endurance by Intermittent Food Deprivation." *FASEB Journal* 32 (2018): 3844–3858.

Mattson, M. P. "Energy Intake and Exercise as Determinants of Brain Health and Vulnerability to Injury and Disease." *Cell Metabolism* 16 (2012): 706–722.

Mattson, M. P., and E. J. Calabrese. *Hormesis: A Revolution in Biology, Toxicology, and Medicine.* New York: Springer, 2010.

Mattson, M. P., K. Moehl, N. Ghena, M. Schmaedick, and A. Cheng. "Intermittent Metabolic Switching, Neuroplasticity, and Brain Health." *Nature Reviews Neuroscience* 19 (2018): 63–80.

Sleiman, S. F., J. Henry, R. Al-Haddad, L. El Hayek, E. Abou Haidar, T. Stringer, D. Ulja, et al. "Exercise Promotes the Expression of Brain Derived Neurotrophic Factor (BDNF) through the Action of the Ketone Body β-hydroxybutyrate." *Elife*, June 2, 2016, 5;e15092.

Stranahan, A. M., and M. P. Mattson. "Recruiting Adaptive Cellular Stress Responses for Successful Brain Ageing." *Nature Reviews Neuroscience* 13 (2012): 209–216.

Sun, N., R. J. Youle, and T. Finkel. "The Mitochondrial Basis of Aging." *Molecular Cell* 61 (2016): 654–666.

Tinsley, G. M., M. Lane Moore, A. J. Graybeal, A. Paoli, Y. Kim, J. U. Gonzales, J. R. Harry, et al. "Time-Restricted Feeding Plus Resistance Training in Active Females: A Randomized Trial." *American Journal of Clinical Nutrition* 110, no. 3 (2019): 628–640.

Van Praag, H., G. Kempermann, and F. H. Gage. "Running Increases Cell Proliferation and Neurogenesis in the Adult Mouse Dentate Gyrus." *Nature Neuroscience* 2, no. 3 (1999): 266–270.

Wan, R., S. Camandola, and M. P. Mattson. "Intermittent Food Deprivation Improves Cardiovascular and Neuroendocrine Responses to Stress in Rats." *Journal of Nutrition* 133, no. 6 (2003): 1921–1929.

Wan, R., L. A. Weigand, R. Bateman, K. Griffioen, D. Mendelowitz, and M. P. Mattson. "Evidence That BDNF Regulates Heart Rate by a Mechanism Involving Increased Brainstem Parasympathetic Neuron Excitability." *Journal of Neurochemistry* 129 (2014): 573–580.

Wilhelmi de Toledo, F., F. Grundler, A. Bergouignan, S. Drinda, and A. Michalsen. "Safety, Health Improvement, and Well-Being during a 4 to 21-Day Fasting Period in an Observational Study Including 1422 Subjects." *PLoS One* 14 (2019): e0209353.

CHAPTER 5

Cheng, A., J. Wang, N. Ghena, Q. Zhao, I. Perone, T. M. King, R. L. Veech, et al. "SIRT3 Haploinsufficiency Aggravates Loss of GABAergic Interneurons and Neuronal Network Hyperexcitability in an Alzheimer's Disease Model." *Journal of Neuroscience* 40 (2020): 694–709.

Clarke, K., K. Tchabanenko, R. Pawlosky, E. Carter, M. T. King, K. Musa-Veloso, M. Ho, et al. "Kinetics, Safety and Tolerability of (R)-3-hydroxybutyl (R)-3-hydroxybutyrate in Healthy Adult Subjects." *Regulatory Toxicology and Pharmacology* 63, no. 3 (2012): 401–408.

Cox, P. J., T. Kirk, T. Ashmore, K. Willerton, R. Evans, A. Smith, A. J. Murray, et al. "Nutritional Ketosis Alters Fuel Preference and Thereby Endurance Performance in Athletes." *Cell Metabolism* 24 (2016): 256–268.

Cunnane, S. C., E. Trushina, C. Morland, A. Prigione, G. Casadesus, Z. B. Andrews, M. Flint Beal, et al. "Brain Energy Rescue: An Emerging Therapeutic Concept for Neurodegenerative Disorders of Aging." *Nature Reviews Drug Discovery* 19, no. 9 (Sept. 2020): 609–633.

Kashiwaya, Y., C. Bergman, J. H. Lee, R. Wan, M. T. King, M. R. Mughal, E. Okun, et al. "A Ketone Ester Diet Exhibits Anxiolytic and Cognition-Sparing Properties, and Lessens Amyloid and Tau Pathologies in a Mouse Model of Alzheimer's Disease." *Neurobiology of Aging* 34, no. 6 (2013): 1530–1539.

Mujica-Parodi, L. R., A. Amgalan, S. F. Sultan, B. Antal, X. Sun, S. Skiena, A. Lithen, et al. "Diet Modulates Brain Network Stability, a Biomarker for Brain Aging, in Young Adults." *Proceedings of the National Academy of Sciences USA* 117 (2020): 6170–6177.

Murray, A. J., N. S. Knight, M. A. Cole, L. E. Cochlin, E. Carter, K. Tchabanenko, T. Pichulik, et al. "Novel Ketone Diet Enhances Physical and Cognitive Performance." *FASEB Journal* 30 (2016): 4021–4032.

CHAPTER 6

Brown, A. W., M. M. Bohan Brown, and D. B. Allison. "Belief beyond the Evidence: Using the Proposed Effect of Breakfast on Obesity to Show Two Practices That Distort Scientific Evidence." *American Journal of Clinical Nutrition* 98, no. 5 (2013): 1298–1308.

Buettner, D. *The Blue Zones: Lessons for Living Longer from People Who've Lived the Longest.* 2nd ed. Washington, DC: National Geographic Society, 2012.

Butler, M., V. A. Nelson, H. Davila, E. Ratner, H. A. Fink, L. S. Hemmy, J. R. McCarten, et al. "Over-the-Counter Supplement Interventions to Prevent Cognitive Decline, Mild Cognitive Impairment, and Clinical Alzheimer-Type Dementia: A Systematic Review." *Annals of Internal Medicine* 168, no. 1 (2018): 52–62.

Camandola, S., N. Plick, and M. P. Mattson. "Impact of Coffee and Cacao Purine Metabolites on Neuroplasticity and Neurodegenerative Disease." *Neurochemical Research* 44, no. 1 (2019): 214–227.

Cheke, L. G., J. S. Simons, and N. S. Clayton. "Higher Body Mass Index Is Associated with Episodic Memory Deficits in Young Adults." *Quarterly Journal of Experimental Psychology* 69, no. 11 (2016): 2305–2316.

Denis, I., B. Potier, S. Vancassel, C. Heberden, and M. Lavialle. "Omega-3 Fatty Acids and Brain Resistance to Ageing and Stress: Body of Evidence and Possible Mechanisms." *Ageing Research Reviews* 12, no. 2 (2013): 579–594.

Dinu, M., G. Pagliai, A. Casini, and F. Sofi. "Mediterranean Diet and Multiple Health Outcomes: An Umbrella Review of Meta-analyses of Observational Studies and Randomised Trials." *European Journal of Clinical Nutrition* 72, no. 1 (2018): 30–43.

Koul, O. *Insect Antifeedants*. New York: CRC Press, 2005.

Lee, J., D. G. Jo, D. Park, H. Y. Chung, and M. P. Mattson. "Adaptive Cellular Stress Pathways as Therapeutic Targets of Dietary Phytochemicals: Focus on the Nervous System." *Pharmacological Reviews* 66, no. 3 (2014): 815–868.

Mattson, M. P., ed. *Diet–Brain Connections: Impact on Memory, Mood, Aging, and Disease*. Norwell, MA: Kluwer Academic, 2002.

Mattson, M. P. "What Doesn't Kill You . . ." *Scientific American* 313, no. 1 (2015): 40–45.

McMacken, M., and S. Shah. "A Plant-Based Diet for the Prevention and Treatment of Type 2 Diabetes." *Journal of Geriatric Cardiology* 14, no. 5 (2017): 342–354.

Noble, E. E., T. M. Hsu, J. Liang, and S. E. Kanoski. "Early-Life Sugar Consumption Has Long-Term Negative Effects on Memory Function in Male Rats." *Nutritional Neuroscience* 22, no. 4 (2019): 273–283.

Ornish, D., L. W. Scherwitz, J. H. Billings, S. E. Brown, K. L. Gould, T. A. Merritt, S. Sparler, et al. "Intensive Lifestyle Changes for Reversal of Coronary Heart Disease." *JAMA* 280, no. 23 (1998): 2001–2007.

Sano, M., C. Ernesto, R. G. Thomas, M. R. Klauber, K. Schafer, M. Grundman, P. Woodbury, et al. "A Controlled Trial of Selegiline, Alpha-Tocopherol, or Both as Treatment for Alzheimer's Disease. The Alzheimer's Disease Cooperative Study." *New England Journal of Medicine* 336, no. 17 (1997): 1216–1222.

Stranahan, A. M., E. D. Norman, K. Lee, R. G. Cutler, R. S. Telljohann, J. M. Egan, and M. P. Mattson. "Diet-Induced Insulin Resistance Impairs Hippocampal Synaptic Plasticity and Cognition in Middle-Aged Rats." *Hippocampus* 18, no. 11 (2008): 1085–1088.

Taubes, G. *The Case against Sugar*. New York: Knopf, 2016.

Ventura, E. E., J. N. Davis, and M. I. Goran. "Sugar Content of Popular Sweetened Beverages Based on Objective Laboratory Analysis: Focus on Fructose Content." *Obesity* (Silver Spring, MD) 19, no. 4 (2011): 868–874.

CHAPTER 7

Augustijn, M. J. C. M., E. D'Hondt, A. Leemans, L. Van Acker, A. De Guchtenaere, M. Lenoir, F. J. A. Deconinck, and K. Caeyenberghs. "Weight Loss, Behavioral Change, and Structural Neuroplasticity in Children with Obesity through a Multidisciplinary Treatment Program." *Human Brain Mapping* 40 (2019): 137–150.

Baker, K. D., A. Loughman, S. J. Spencer, and A. C. Reichelt. "The Impact of Obesity and Hypercaloric Diet Consumption on Anxiety and Emotional Behavior across the Lifespan." *Neuroscience Biobehavioral Reviews* 83 (2017): 173–182.

Baker, K. D., and A. C. Reichelt. "Impaired Fear Extinction Retention and Increased Anxiety-Like Behaviours Induced by Limited Daily Access to a High-Fat/High-Sugar Diet in Male Rats: Implications for Diet-Induced Prefrontal Cortex Dysregulation." *Neurobiology of Learning and Memory* 136 (2016): 127–138.

Bustamante, E. E., C. F. Williams, and C. L. Davis. "Physical Activity Interventions for Neurocognitive and Academic Performance in Overweight and Obese Youth: A Systematic Review." *Pediatric Clinics of North America* 63 (2016): 459–480.

Cao-Lei, L., D. P. Laplante, and S. King. "Prenatal Maternal Stress and Epigenetics: Review of the Human Research." *Current Molecular Biology Reports* 2 (2016): 16–25.

CDC. "Data & Statistics on Autism Spectrum Disorder." Centers for Disease Control and Prevention. n.d. Accessed April 21, 2021. https://www.cdc.gov/ncbddd/autism/data.html.

De Luca, S. N., I. Ziko, L. Sominsky, J. C. Nguyen, T. Dinan, A. A. Miller, T. A. Jenkins, and S. F. Spencer. "Early Life Overfeeding Impairs Spatial Memory Performance by Reducing Microglial Sensitivity to Learning." *Journal of Neuroinflammation* 13 (2016): 112.

Ferreira, A., J. P. Castro, J. P. Andrade, M. Dulce Madeira, and A. Cardoso. "Cafeteria-Diet Effects on Cognitive Functions, Anxiety, Fear Response and Neurogenesis in the Juvenile Rat." *Neurobiology of Learning and Memory* 155 (2018): 197–207.

Gray, J. C., N. A. Schvey, and M. Tanofsky-Kraff. "Demographic, Psychological, Behavioral, and Cognitive Correlates of BMI in Youth: Findings from the Adolescent Brain Cognitive Development (ABCD) Study." *Psychology and Medicine*, July 10, 2019, 1–9.

Mattson, M. P. "An Evolutionary Perspective on Why Food Overconsumption Impairs Cognition." *Trends in Cognitive Sciences* 23, no. 3 (2019): 200–212.

Phillips, O. R., A. K. Onopa, Y. V. Zaiko, and M. K. Singh. "Insulin Resistance Is Associated with Smaller Brain Volumes in a Preliminary Study of Depressed and Obese Children." *Pediatric Diabetes* 19 (2018): 892–897.

Rivell, A., and M. P. Mattson. "Intergenerational Metabolic Syndrome and Neuronal Network Hyperexcitability in Autism." *Trends in Neurosciences* 42 (2019): 709–726.

Ross, N., P. L. Yau, and A. Convit. "Obesity, Fitness, and Brain Integrity in Adolescence." *Appetite* 93 (2015): 44–50.

Weber, A. S., ed. *Nineteenth-Century Science: An Anthology.* Peterborough, CA: Broadview Press, 2000.

Yau, P. L., M. G. Castro, A. Tagani, W. H. Tsiu, and A. Convit. "Obesity and Metabolic Syndrome and Functional and Structural Brain Impairments in Adolescence." *Pediatrics* 130 (2012): e856–e864.

Yau, P. L., E. H. Kang, D. C. Javier, and A. Convit. "Preliminary Evidence of Cognitive and Brain Abnormalities in Uncomplicated Adolescent Obesity." *Obesity* (Silver Spring, MD) 22 (2014): 1865–1871.

CHAPTER 8

Athauda, D., K. Maclagan, S. S. Skene, M. Bajwa-Joseph, D. Letchford, K. Chowdhury, S. Hibbert, et al. "Exenatide Once Weekly versus Placebo in Parkinson's Disease: A Randomised, Double-Blind, Placebo-Controlled Trial." *Lancet* 390 (2017): 1664–1675.

Bradley, E. H., and L. A. Taylor. *The American Health Care Paradox: Why Spending More Is Getting Us Less.* New York: PublicAffairs, Perseus Books Group, 2013.

Brill, S. *America's Pill: Money, Politics, Backroom Deals, and the Fight to Fix Our Broken Healthcare System.* New York: Random House, 2015.

Doyle, M. E., P. McConville, M. J. Theodorakis, M. M. Goetschkes, M. Bernier, R. G. S. Spencer, H. W. Holloway, et al. "In Vivo Biological Activity of Exendin (1-30)." *Endocrine* 27, no. 1 (2005): 1–9.

Duan, W., and M. P. Mattson. "Dietary Restriction and 2-Deoxyglucose Administration Improve Behavioral Outcome and Reduce Degeneration of Dopaminergic

Neurons in Models of Parkinson's Disease." *Journal of Neuroscience Research* 57 (1999): 195–206.

Emond, A. J., A. M. Bernhardt, D. Gilbert-Diamond, Z. Li, and J. D. Sargent. "Commercial Television Exposure, Fast Food Toy Collecting, and Family Visits to Fast Food Restaurants among Families Living in Rural Communities." *Journal of Pediatrics* 168 (2016): 158–163.e1.

Geisler, J. G., K. Marosi, J. Halpern, and M. P. Mattson. "DNP, Mitochondrial Uncoupling, and Neuroprotection: A Little Dab'll Do Ya." *Alzheimer's and Dementia* 13 (2017): 582–591.

Harrison, D. E., R. Strong, Z. D. Sharp, J. F. Nelson, C. M. Astle, K. Flurkey, N. L. Nadon, et al. "Rapamycin Fed Late in Life Extends Lifespan in Genetically Heterogeneous Mice." *Nature* 460 (2009): 392–395.

Kim, W., and J. M. Egan. "The Role of Incretins in Glucose Homeostasis and Diabetes Treatment." *Pharmacological Reviews* 60 (2008): 470–512.

Kishimoto, Y., J. Johnson, W. Fang, J. Halpern, K. Marosi, D. Liu, J. G. Geisler, et al. "A Mitochondrial Uncoupler Prodrug Protects Dopaminergic Neurons and Improves Functional Outcome in a Mouse Model of Parkinson's Disease." *Neurobiology of Aging* 85 (2020): 123–130.

Lautrup, S., D. A. Sinclair, M. P. Mattson, and E. F. Fang. "NAD$^+$ in Brain Aging and Neurodegenerative Disorders." *Cell Metabolism* 30 (2019): 630–655.

Lee, J., A. J. Bruce-Keeler, Y. Kruman, S. L. Chan, and M. P. Mattson. "2-Deoxy-D-Glucose Protects Hippocampal Neurons against Excitotoxic and Oxidative Injury: Evidence for the Involvement of Stress Proteins." *Journal of Neuroscience Research* 57, no. 1 (1999): 48–61.

Li, Y., T. A. Perry, M. S. Kindy, B. K. Harvey, D. Tweedie, H. W. Holloway, K. Powers, et al. "GLP-1 Receptor Stimulation Preserves Primary Cortical and Dopaminergic Neurons in Cellular and Rodent Models of Stroke and Parkinsonism." *Proceedings of the National Academy of Sciences USA* 106 (2009): 1285–1290.

Liu, D., M. Pitta, H. Jiang, J-H. Lee, G. Zhang, X. Chen, E. M. Kawamoto, and M. P. Mattson. "Nicotinamide Forestalls Pathology and Cognitive Decline in Alzheimer Mice: Evidence for Improved Neuronal Bioenergetics and Autophagy Procession." *Neurobiology Aging* 34 (2013): 1564–1580.

Liu, D., Y. Zhang, R. Gharavi, H. R. Park, J. Lee, S. Siddiqui, R. Telljohann, et al. "The Mitochondrial Uncoupler DNP Triggers Brain Cell mTOR Signaling

Network Reprogramming and CREB Pathway Up-Regulation." *Journal of Neurochemistry* 134, no. 4 (2015): 677–692.

Malagelada, C., Z. H. Jin, V. Jackson-Lewis, S. Przedborski, and L. A. Greene. "Rapamycin Protects against Neuron Death in In Vitro and In Vivo Models of Parkinson's Disease." *Journal of Neuroscience* 30 (2010): 1166–1175.

Otto, S. *The War on Science: Who's Waging It, Why It Matters, What We Can Do about It.* Minneapolis, MN: Milkweed, 2016.

Rajman, L., K. Chwalek, and D. A. Sinclair. "Therapeutic Potential of NAD-Boosting Molecules: The In Vivo Evidence." *Cell Metabolism* 27 (2018): 529–547.

Rosenthal, E. *An American Sickness: How Healthcare Became Big Business and How You Can Take It Back.* New York: Penguin Books, 2017.

Sadeghirad, B. "Influence of Unhealthy Food and Beverage Marketing on Children's Dietary Intake and Preference: A Systematic Review and Meta-analysis of Randomized Trials." *Obesity Reviews* 17 (2016): 945–959.

Salcedo, I., D. Tweedie, Y. Li, and N. H. Greig. "Neuroprotective and Neurotrophic Actions of Glucagon-Like Peptide-1: An Emerging Opportunity to Treat Neurodegenerative and Cerebrovascular Disorders." *British Journal of Pharmacology* 166, no. 5 (2012): 1586–1599.

Volkow, N. D., R. A. Wise, and R. Baler. "The Dopamine Motive System: Implications for Drug and Food Addiction." *Nature Reviews Neuroscience* 18, no. 12 (2017): 741–752.

Wu, T. *The Attention Merchants: The Epic Scramble to Get Inside Our Heads.* New York: Knopf, 2016.

Yu, Z. F., and M. P. Mattson. "Dietary Restriction and 2-Deoxyglucose Administration Reduce Focal Ischemic Brain Damage and Improve Behavioral Outcome: Evidence for a Preconditioning Mechanism." *Journal of Neuroscience Research* 57, no. 6 (1999): 830–839.

CHAPTER 9

Cunnane, S. C., E. Trushina, C. Morland, A. Prigione, G. Casadesus, Z. B. Andrews, M. Flint Beal, et al. "Brain Energy Rescue: An Emerging Therapeutic Concept for Neurodegenerative Disorders of Aging." *Nature Reviews Drug Discovery* 19, no. 9 (Sept. 2020): 609–633.

De Cabo, R., and M. P. Mattson. "Impact of Intermittent Fasting on Health, Aging, and Disease." *New England Journal of Medicne* 381, no. 26 (Dec. 2019): 2541–2551.

Harvie, M. N., M. Pegington, M. P. Mattson, J. Frystyk, B. Dillon, G. Evans, J. Cuzick, et al. "The Effects of Intermittent or Continuous Energy Restriction on Weight Loss and Metabolic Disease Risk Markers: A Randomized Trial in Young Overweight Women." *International Journal of Obesity* (London) 35, no. 5 (2011): 714–727.

Rose, S. *The Future of the Brain: The Promise and Perils of Tomorrow's Neuroscience.* Oxford: Oxford University Press, 2005.

Index